STAR-CRAVING MAD

FRED WATSON

STAR-CRAVING MAD

TALES FROM A TRAVELLING ASTRONOMER

ALLEN&UNWIN

First published in 2013

Allen & Unwin
Sydney, Melbourne, Auckland, London

83 Alexander Street
Crows Nest NSW 2065
Australia
Phone: (61 2) 8425 0100
Email: info@allenandunwin.com
Web: www.allenandunwin.com

Cataloguing-in-Publication details are available
from the National Library of Australia
www.trove.nla.gov.au

ISBN 978 1 74237 376 8

Index by Penny Mansley
Set in 11/13.5 pt Janson Text by Post Pre-press Group, Australia
Printed in Australia by McPherson's Printing Group

10 9 8 7 6 5 4 3 2 1

To my marvellous modern family,
with love

CONTENTS

ACKNOWLEDGEMENTS

Years ago, I had a letter from an Italian gentleman, who was grateful for some publications I had sent him from the Royal Observatory, Edinburgh. Written in less-than-perfect English, the letter ended with the memorable line 'Thank you for your remarkable disposability'. While I *think* I know what he meant, I've never been quite sure. Far less disposable than me, however, are the many people who have contributed to this book, and it is a great pleasure to have the opportunity to thank them.

None of the study tours described here would have been possible without the consummate expertise of my partner and manager, Marnie Ogg, so my first and biggest thank-you goes to her. Turning such trips into reality requires an army of tour companies, tour directors and local guides, as well as a deep understanding of the travel business, and Marnie has all that and more at her fingertips. It would be hard to overstate her contribution to this book—not the least being its title. I'd also like to thank my fellow travellers on the tours for their enthusiastic participation and great company. I think I've learned as much from them as they have from me.

Then there is the organisation that keeps me going, and it is a pleasure to acknowledge the support of the Australian Astronomical Observatory, a division of the Commonwealth Department of Industry, Innovation, Science, Research and Tertiary Education. I'm especially indebted to the Director, Matthew Colless, for his constant encouragement, and helpful comments on some chapters of the book. Many thanks, too, to Neville Legg, General Manager, for always checking that my recreation leave is in order before I go on tour—and much more. The friendly support of everyone on the observatory's staff, both in Sydney and at Siding Spring, is gratefully acknowledged.

The study tours themselves have benefited enormously from the generosity of colleagues the world over in lending their expertise during our visits. I thank Mattias Abrahamsson, Bob Argyle, Klaus Bätzner, John Brown, Andrew Collier Cameron, Iván Ghezzi, Ann-Christin Grenevall, Mark Hurn, Andrew Jacob, Lennart Jonasson, Quentin King, Michael Linden-Voenle, Nick Lomb, Andy Longmore, Peter Louwman, Karen Moran, Ulisse Munari, Pasi Nurmi, Nick Petford, Bertil Pettersson, Dominique Proust, Alan Pickup, Rami Rekola, John Sarkissian, Felix and Susanne Seiler, Urmas Sisask, Alessandro Siviero, Matthias Steinmetz, Toner Stevenson and Geoff Wyatt.

I owe a special debt to three distinguished St Andrews graduates: Þorsteinn Sæmundsson, who shared his personal recollections of Erwin Finlay Freundlich with me during an unforgettable visit to Iceland; our mutual friend, Bob Shobbrook, who put us into contact; and Edmund Robertson, of the superb *MacTutor History of Mathematics Archive*, for his insights into the Freundlich era at St Andrews. It's also a great pleasure to acknowledge Howard Sacre, of Nine Network Australia, for masterminding the documentary filming at the Large Hadron Collider. Likewise, my gratitude to my co-conspirator on the show, Liam Bartlett.

The historical accounts in *Star-Craving Mad* draw on the published work of Peter Aughton, André Baranne, Jonas Bendiksen, J.A. Bennett, Geoffrey Blainey, Terrie F. Bloom, Randall C. Brooks, Allan Chapman, John R. Christianson, Tom Frame, Don Faulkner, the late Ben Gascoigne, Owen Gingerich, Ian Glass, W. Gratzner, Richard F. Harrison, John B. Hearnshaw, Michael Hoskin, Stefan Ilsemann, Lucy Jago, the late Henry C. King, Kenneth R. Lang, Françoise Launay, Juan Carlos Machicado Figueroa, J.P. McEvoy, John J. O'Connor, the late M. Barlow Pepin, Katrina Proust, M.O. Robins, Andrew Robinson, the late Colin A. Ronan, Clive L.N. Ruggles, Alan D.C. Simpson, Engel Sluiter, the late Victor E. Thoren, A.J. Turner, Albert Van Helden, Arne Wennberg, Richard S. Westfall, Robert S. Westman and Rolf Willach. To these accomplished historians, I express my admiration and gratitude.

As regards the scientific content of the book, it's hard to know where to start in acknowledging all the friends and colleagues who have provided input over the years. But I'd particularly like to mention Peter Abrahams, Jeremy Bailey, Tim Beers, Brian Boyle, Russell Cannon, Brad Carter, Paul Cass, Victor Clube, Warrick Couch, Phil Diamond, Roger Davies, Peter Downes, Ken Freeman, Gerry Gilmore, Peter Gray, Malcolm Hartley, Joss Hawthorn, Rob Hollow, Andrew Hopkins, Stephen Hughes, Chris Impey, Hugh Jones, Dennis Kelly, David Kilkenny, John Lattanzio, Charley Lineweaver, Malcolm Longair, David Malin, John Mason, Rob McNaught,

Patrick Moore, Ray Norris, Simon O'Toole, Quentin Parker, John Peacock, Mike Read, Ken Russell, Stuart Ryder, Brian Schmidt, Milorad Stupar, Chris Tinney, Pete Wheeler, Doug Whittet, Reg Wilson, Joe Wolfe and Tomaz Zwitter. I also acknowledge the role of my colleagues in the galactic archaeology community and the RAVE consortium in the work reported in Chapter 7.

Other friends who have enthusiastically supported my efforts in science outreach include John Budge, Donna Burton, Marcus Chown, Antony Cooke, Rob Dean, Rosalind Dubs, Ross and Helen Edwards, Ron Ellis and Susan Murray, Kristin Fiegert, Hans and Frances Gnodtke, Doug Gray, Derrick and Lorna Hartley, Laura Hartley, Ray and Libby Johnson, Phillipa Malin, Haritina Mogosanu, Jeff and Dianne Ogg, Matthew Ogg and Mirjam Beck, Robyn Owens, Sue Rawlings, William and Nina Reid, Victor and Sandra Richardson, Helen Sim, Peter Slezak, Dava Sobel, Michael Sollis and the Griffyn Ensemble, Colin and Anne Spencer, and Robyn Williams. Not to mention a lot of people at the ABC, Nine Network Australia and Network Ten.

Star-Craving Mad owes its origin to Ian Bowring of Allen & Unwin, but I took so long over the project that he retired in the meantime. Gosh, Ian, where's your staying-power? Thankfully, the reins were taken over by Foong Ling Kong, and the book has been marvellously edited by Ann Lennox, Penny Mansley and Susan Jarvis. Grateful thanks to them all.

And so to my family. I thank Alan and Monica Watson and David Garnett for sharing their knowledge of family history. Thanks, too, to my brother, John Watson, and my son James for checking over the final chapters. Formalities aside, my four children—Helen, Anna, James and Will—make me very proud; all the more so now that the girls have beautiful families of their own. Their continuing support is always cherished. And, finally, the person who masterminded the tours in this book is also the person who is my shining light in the world. Not a day goes by without me feeling truly grateful for her sparkling presence. Thanks for everything, Marnie.

1 INTERPLANETARY TRAVELLERS

Journeys through space and time

Have you ever met anyone from Pluto? I have—or, at least, that's where he said he was from when I met him. He was very striking: tall, dreadlocks, a vivid-pink silk suit, and carrying something that looked at first sight like a didjeridu. Since this was Berlin, that seemed unlikely, and, indeed, it did turn out to be nothing more than a big stick. It was the kind of thing you might take to a fancy dress party if you went along as a prophet. So I guess it should have come as no surprise that this gentleman eventually revealed that he was, well, a prophet.

He had been sitting with a couple of friends—disciples, perhaps—in the back row of a small lecture theatre in the Urania science centre, where I'd been giving a talk about Pluto to an audience of science-minded Aussie travellers and curious Berliners. The

Aussie travellers had just joined me for a study tour of Europe, while the Berliners might only have been there for a bit of English language practice. Who knows? It always pays to keep an eye on the back row, since this is traditionally where the naughty seats are, and old habits die hard—even among otherwise responsible adults. But the pink gentleman had looked harmless enough, if a little eccentric, so when he rose with messianic import in response to my invitation for questions, I was a bit taken aback.

'Yes, Professor Watson,' he began, in an alarmingly authoritative tone. That took me aback, too, since most members of the public who come to my talks have no idea what my name is, and if they do they just call me Fred. Which I much prefer. 'Professor Watson, I have the first question.'

OK, here it comes, I thought. He's going to take issue with me over Pluto's relegation to a dwarf planet. That wasn't something I was in any way responsible for, but I'd made much of it in my discourse, as I had tried to explain the scientific rationale behind it.

But no, it wasn't that. 'I would like to know which direction Pluto is standing in at the present time.'

Well, I guess it was a fair question, given the subject of the lecture. But it's not the sort of thing most astronomers carry around in their heads. The pink gentleman elaborated a bit by explaining that he wanted its position in degrees, and I suppose I could have told him that Pluto was 21 years past its perihelion point and invited him to calculate its celestial longitude. In degrees. In his head. However, since I couldn't have done it myself, he could have given me any number he liked. But I did tell him how to find the answer—by consulting the truly wonderful Heavens Above website, which gives up-to-the-minute celestial positions for all the main bodies of the Solar System. That seemed to be acceptable.

'OK, next question,' he went on. 'Can you tell me if there is any scientist on the face of this Earth who can give me the exact distance from the Earth to Pluto?'

When I replied that we probably know Pluto's distance with an accuracy of a few kilometres—which is not bad for something that's nearly 5 billion kilometres away—he looked singularly unimpressed. So did his disciples, who had clearly been expecting something more entertaining than a discussion about mere Solar System distances.

It seemed it was time for him to throw down the gauntlet. 'That's really just speculation, though, isn't it?' he said.

Well, I would have thought that a few kilometres in 5 billion kilometres was pretty good speculation, so I began to explain how we know the distance.

But that wasn't in the pink gentleman's script. 'I'd like to introduce myself. My name is Messenger Nine, from the School of Prophets, sent to you by my master, Pluto.' Ah, this was better. The disciples looked relieved. So did everyone else in the room, but probably for a different reason. At least they now knew what they were up against. 'At the School of Prophets, we don't deal in speculation—only certainties.' The disciples beamed. 'And I have a prophecy to make.' His tone became more messianic by the minute as he brought us his forecast of doom. 'I was sent here to prophesy to this scientific community about an event that will unfold in seven days. It will be an earthquake, of magnitude seven or eight, and it will be caused by the planet Pluto.' By now, Messenger Nine was in full flow, and the disciples were beside themselves with admiration.

I ventured to interrupt. 'So can you tell me where this earthquake will occur?'

'Well, let me tell you,' he began again, drawing breath for another assault on his disbelieving audience.

But the polite German chair of the session had sensed that Messenger Nine was about to embark on an answer that would ramble far and wide, and interrupted him. 'Excuse me. It's his talk, not yours.'

Laughter and cheers from the audience.

'OK. Well, let me tell you it will occur two hundred and forty-three degrees west of the equator.'

I don't think Messenger Nine had bargained for the tirade of protest he received from the audience, who were all awake enough to realise that 243 degrees west of the equator is meaningless as a position on the Earth's surface. It's like saying 'Ten kilometres along the Sydney road.' Ten kilometres from where? He stuck to his guns, though, no doubt for the benefit of the disciples, and seemed convinced that such an earthquake would occur. Eventually, amid growing protests from the audience, Messenger Nine was persuaded to sit down so that others could ask questions, which, in comparison with his, were perfectly normal.

Once the formalities of the evening were over and people were beginning to drift away, Messenger Nine wandered up to the front of the lecture theatre for a bit of a chat. He admitted that he wasn't actually from Pluto but from Georgetown, Guyana. Pluto was more of a spiritual home, you see.

I didn't discover what he was doing in Berlin, but I rather liked him and admired his pink outfit and stick. So I agreed to give him a call in the event of an earthquake occurring anywhere in the world within a week. And, since that undertaking was witnessed by a dozen or so folk who would be my fellow travellers for the next fortnight, I was duty-bound to keep my word. But there was no earthquake that week. Not even a modest crockery-rattler.

The best thing about Messenger Nine was that he got our 2010 Stargazer II tour of Europe off to a fine start. His pink suit was a great talking point and proved the perfect ice-breaker for a group of people who, while they had similar interests and enthusiasms, for the most part didn't know one another. Some tour members even accused me of planting him in the audience to entertain the punters. While that was quite a good idea, I swear it wasn't what had taken place. But what were we doing in Berlin? And why was I giving a talk there about Pluto?

ASTRONOMY TOURISM

The answers to those questions go back a long way. Once upon a time, I was a shop-floor astronomer quietly going about my business trying to solve obscure mysteries concerning our Milky Way Galaxy—the Sun's home in the Universe, which it shares with 400 billion or so other stars. My work was (and, in fact, still is) to concentrate on just one tiny part of the giant jigsaw puzzle of knowledge that scientists have built up about our wider environment.

It's surprising how many people outside the science world harbour the romantic notion that astronomers spend every night with their eyes glued to giant telescopes, looking for things. Just, well, looking. The most frequent question I'm asked by members of the public is 'Have you found anything recently?' Sadly, apart from the odd sock that has made a bid for freedom from the washing basket, the answer is usually 'No.' Finding new things is only a small part of what we do, compared with investigating things we already know about. Well-defined research targeting particular questions is the name of the game. And today we rely on advanced technology to do that—meaning that computer screens and hard drives have long replaced telescope eyepieces as the astronomer's window on the Universe.

Where's the romance in that, then? Apart from the enticing prospect of a discovery, the romance lies in the nature of the investigations we carry out. Astronomy is basically large-scale sleuthing—electronic eavesdropping on events in deep space to address the big questions: 'Where did we come from?' 'How did our planet get to be the way it is?' 'How did the Universe get to be the way it is?' 'What will happen to it in the end?' And, perhaps the biggest question of all: 'Are we alone, or are there other living organisms beyond the Earth?' The day-to-day work of an astronomer is, however, remarkably unglamorous and often far removed from the lofty aims of the science. The routine tasks of telescope observing, data analysis, paper writing, seminar speaking, grant applications and all the rest often seem very humdrum indeed. Particularly if you struggle a bit with some of them, as I do. But one thing that has always motivated me strongly is the thought that, ultimately, this research is being paid for by the man and woman in the street—the nation's taxpayers. And these folk are generally interested in the outcomes. Perhaps surprisingly, their interest is not driven by questions about where their hard-earned money is going, but by pure curiosity.

In Australia, at least, it is extraordinary what a healthy appetite the general public has for astronomy and space science. People sense that in this biggest of big sciences there might be answers to some of the most profound questions we can ask. Questions about the nature of space and time, about our ultimate origins, the meaning of life and perhaps even spirituality—although I find it's safer to keep God Herself out of the equation. As my erstwhile PhD supervisor used to remind me, astronomy doesn't tell you about God; it tells you about the Universe. Nevertheless, astronomy does provide a broader framework than most sciences for deliberations about such profound issues, as we shall see later in this book.

But the public's interest doesn't stop at the big questions. People notice all sorts of little things, like rings around the Sun and Moon, groupings of planets in the sky, gradual changes in sunrise and sunset times, and fly-bys of the International Space Station. There's almost nothing within what astronomers might regard as commonplace that doesn't fascinate the public at large.

Best of all, the public appreciates that, as a science, astronomy is generally above reproach. Since there is no marketable end-product, there is little scope for corruption, either by wheeling and dealing or by anything that would disadvantage a particular group of individuals. Nor is there any immediate application for astronomy in defence or politics. You could almost describe it as the 'honest broker of the sciences'. Why? Because no science could be more honest and, well, no science could be broker. (Even though astronomy is reasonably well supported in most developed countries, the funding is never lavish.)

In the face of such overwhelming interest, how could any astronomer fail to engage with the wider public? So it was that, many years ago, I became involved in science outreach. It started with popular-level talks and magazine articles and then blossomed into broadcasting (despite my first attempt being dumped by the BBC, just because Argentina chose the same morning to invade the Falkland Islands). Eventually, I started writing astronomy books, and a lifelong fascination with the evolution of astronomy, from the earliest primitive musings about the sky to our present state of knowledge, broke through to the surface. And an exciting new topic suggested itself: the geography of astronomy.

Astronomy has, over the ages, established centres of excellence where great discoveries or advances have been made. Many of them still exist, and they are often in rather beautiful parts of the world. There are also places on Earth that are intimately associated with astronomical research being carried out today. One only has to think of Coonabarabran and Parkes in Australia—home of the Anglo-Australian Telescope and the Parkes Radio Dish respectively—to gain a sense of these iconic scientific locations. Our planet is full of such places, ranging from remote, inhospitable mountain tops where great telescopes ply their trade to the leafy suburbs of cities like Geneva, home of Europe's Large Hadron Collider. As well as being our spaceship through the Universe, the Earth is also our observatory and laboratory, with great scientific centres of the past and present available to be experienced at first hand: they are there for the visiting.

My first experience of a real astronomical observatory took place a long time ago—a very long time ago, in fact, when I was a spotty-faced teenager. My dad took me to an old manor house not far from our home in the drab, industrial north of England. This building—Horton Hall—had once been an imposing residence set in open countryside, but it was now empty and dilapidated, and surrounded by suburban housing and light industry. It was close to where Dad worked, and he had guessed that its run-down state made it ripe for demolition. Indeed, that's exactly what happened soon afterwards, to the everlasting shame of the local authority.

'So,' said Dad when we arrived, 'what do you think this place was?'

'Don't know.' (Remember, I was a teenager.)

'Any guesses? Does the tower in the middle give you a clue?'

Classic teenage shrug of the shoulders. 'Don't know. A lighthouse?'

'Freddy, we're sixty miles from the sea.'

'Yes, I know, but . . .' (Another shrug.) 'Well, was it a factory?'

'Freddy, it's three hundred years old, you gormless ha'porth.'

And so on. But it wasn't long before my kind-hearted dad revealed why he had taken me there. It was all about my growing obsession with astronomy and space science. Dad had heard that two and a half centuries earlier this had been the home of a famous astronomer, a fellow Yorkshireman, by the name of Abraham Sharp, who had worked with England's first astronomer royal. Its now-crumbling central tower, crowned with a balustraded platform, had formed his observatory, from which he could gaze at the sky through the crude, stick-like telescopes of the day.

I don't remember many details of our impromptu exploration of this dusty old building, except that it was fraught with all the dangers associated with a forthcoming demolition site. I do recall the tower being out of bounds. Even so, by today's standards, we broke every occupational health and safety rule in the book. Was I scared? Not a bit.

I was enchanted. Being inside this building was the nearest I had ever come to real astronomy, and, I can tell you, it was spine-tingling stuff for a star-struck kid.

Not many years after that I went off to university and in due course began the random stumble through science that I now cherish as my career. It took me to some of the most famous observatories in the world. But I never forgot the fascination of exploring the place that had once been home to an important astronomer. It was as if experiencing the surroundings that old Abraham Sharp had been familiar with formed a bridge across the centuries, an inspiration to any would-be follower in his footsteps. It seemed to me that this link was something rather precious, a sense heightened in the case of Horton Hall by its location, well off the tourist track, and its sad demise soon after my visit.

When I began to think about astronomy's geography, I was struck by an intriguing question: 'If I could feel the sense of connection offered by bridges across time created by astronomical sites, why shouldn't other folk?' By that time, global travel was easy and, as I've mentioned, there was a great community of interested bystanders who would be up for the trip. Astronomy tourism—what a bonzer idea.

As I began to contemplate the possibilities of astronomy tourism, I was reminded of an unexpected but welcome asset that my job has provided over the years: a vast company of friends in the trade—astronomers at institutions all over the world.

Compared with other sciences, astronomy has relatively few practitioners. There are perhaps only 10 000 or so professional astronomers around the globe. Being part of such a small community, astronomers tend to know their colleagues everywhere, and most of them will happily show off all the hidden treasures of their home institutions at the drop of a hat. Perhaps the most stimulating aspect of astronomy tourism could be that, if it were run by professional astronomers with these contacts, it might present the general public with rare opportunities to interact with the scientists carrying out today's research, whether in the front line of physics, astronomy or astrobiology, or in studies of the legacy of the past. It would help 'normal' people to answer the question 'How do we know what we know about the Universe?'

Astronomers are themselves inveterate travellers, often journeying halfway around the world to make their celestial observations. They have recognised the value of building telescopes on world-class observing sites rather than just wherever they happened to be working. For me, it was a small step from travelling for research to travelling as a leader of study tours. For me, astronomy tourism has become a big thing in recent years.

The delicious blend of people, places, events and scientific insights that I have experienced through astronomy tourism has become an intoxicating brew for me, and it's my fond hope that it will add spice to this book's chapters.

PEOPLE, PLACES AND EVENTS

Who are the astronomy tourists, the people who come along with me on these trips? They are folk just like you, who save up for an annual holiday but instead of following the crowds decide to do something a little different—something that will expand their horizons in unusual ways. They travel in relatively small numbers, perhaps a dozen or so at a time, which provides a special intimacy with the subject matter and often sparks long-lasting friendships. The small numbers also allow them to visit places where bigger groups couldn't go.

The secret of successful astronomy touring is to have an effective tour coordinator, and I was fortunate in having one of those from the outset—someone with a clear understanding not only of what would prove irresistible to science and history buffs but of what would also meet the needs of their sometimes not-so-interested partners. That reflects no particular skill on my part because, in fact, my coordinator found me rather than the other way around. I've also been fortunate in having some good friends from the world of astronomy—people whose skills I respect enormously—to share the load in leading the tours. Friends like David Malin, the world's foremost astronomical photographer, who, in the 1980s, showed us the true colours of the stars for the first time, and Ray Norris, a radio astronomer with profound insights into the Dreamtime sky stories of Australia's Indigenous peoples.

Where have we been? As they used to say in the 1960s, we've been everywhere, man. Well, not quite. In reality, we've only scratched the surface of what is possible. Perhaps the most obvious destinations for astronomy tourists set astronomical events in their context. Given that none of us can visit the planet Uranus, for example, what better alternative than to visit Bath, in Somerset, the place from where Uranus was discovered back in 1781, and, while so doing, to learn what everyday life was like for your average well-to-do Englishman at the time? Or why not return to the infancy of Albert Einstein's General Theory of Relativity and find the strange, Expressionist tower telescope that was built in the early 1920s to prove him correct? Or, coming right up to date, rather than read about a new European radio astronomy array whose individual antennae are spread out like the turbines of a miniature wind farm, why not actually stand in a fragrant French field amid dozens of its carefully planted antenna poles?

There's more to astronomy tourism than merely roaming the world's scientific beauty spots and exploring their hidden treasures, though. Some heavenly spectaculars can be seen only from particular vantage points on the Earth's surface. The Aurora Borealis—the Northern Lights—for example, is best seen from the Arctic Circle. Then there is the Aurora Borealis' southern-hemisphere counterpart, the Aurora Australis. And the most awesome of all celestial phenomena, a total eclipse of the Sun. Astronomy tourism can—and does—cater for those, too.

This brings us to the reason why our little band of Aussie astronomy tourists was in Berlin—and why I was spouting about Pluto to anyone who would listen. Why do people usually visit Berlin? To hear the legendary Berlin Philharmonic Orchestra on its home turf, perhaps. Or to take an insanely dangerous drive through the rush-hour traffic in a Cold War East German Trabi with its smoky two-stroke engine and death-or-glory brakes. And to be stopped while driving said Trabi by an alarmingly realistic member of the East German secret police—the dreaded Stasi—who insists on extracting a kiss from any attractive occupant of your car as the price of continued freedom. Or, more seriously, to gasp at the lengths to which the former Communist regime went in dividing the city and preventing the free passage of its people, and then to be inspired by the splendid reunification of the city's heart at the Brandenburg Gate and the Reichstag. We did all of the above—and more—but none of them was the real reason why we were there.

Berlin is the home of several places of astronomical interest, including a planetarium and some historical observatories. More importantly, it is located within a stone's throw of Potsdam, pearl of the old German Democratic Republic, which boasts one of the most historically significant observatories in northern Europe. Now the home of the Leibniz Institute for Astrophysics, the observatory has two of the greatest refracting (lens) telescopes of the nineteenth century as well as that curious tower telescope I mentioned earlier, the Einstein Tower, at the Telegrafenberg site. Wonderful stuff. But it wouldn't have been quite so wonderful without our trump card. The modern Leibniz Institute is a centre of excellence in European astronomy, and its director is Matthias Steinmetz, who, by dint of a project with the unlikely name of RAVE, is a colleague and friend of mine. RAVE will feature later in this book, but suffice it to say for now that our little group was extraordinarily privileged to be given a specially tailored tour by one of Germany's most distinguished scientists. And it was a blast.

Berlin's connection with Pluto is a bit less clear. For us on the tour, however, Berlin was a stopping-off point for Prague, our next destination, 350 kilometres to the south. And Prague, as we shall see, has a very close connection with Pluto.

TECTONIC TOURISM

But what of Messenger Nine and his earthquake? Some readers may think I simply invented him to spice up the opening of this book, just as some of my tourists believed I'd planted him in the audience of my talk, but I promise I didn't. In fact, rather unexpectedly, the whole encounter with the pink prophet was recorded for posterity by the organisers of the talk. At Berlin's Urania science centre, visiting lecturers are routinely handed a gift-wrapped audio CD of their presentation at the end of the event. Courteous and efficient—but this is Germany, after all. I'm not particularly hooked on keeping records of my public ramblings, but this one did provide me with an account of everything Messenger Nine had to say, as faithfully reported to you a few pages ago. Messenger Nine wasn't a prophet, of course. He was just someone who had gone a bit too far down the rocky road of believing his own hype. It can happen to the best of us. He was certainly one of the most unusual characters I've ever come across at one of my talks. But he inadvertently highlighted a further group of destinations that are of interest to astronomy tourists, and to understand the relevance of these destinations to astronomy and its geography we must recall some ground-shaking events.

It was significantly more than a week after my encounter with Messenger Nine that a magnitude-7 earthquake occurred. Three months and fourteen days, to be exact. When it came, it further defied his prediction by occurring at latitude 43.5 degrees south, and longitude 172.6 degrees east. No one who was in Christchurch, New Zealand, on 4 September 2010 will ever forget that early-morning wake-up call. Of course, we now know that this was just a precursor to the far more devastating earthquake of 22 February 2011, in which 182 people tragically lost their lives. And, barely two weeks later, that too was eclipsed in destructive power. The whole world watched in horror as the Tohoku earthquake and the incredible tsunami that followed reaped wholesale devastation on the Japanese coast. The most recent analysis of that dreadful event suggests that in places the tsunami reached a height of over 40 metres—as high as a fifteen-storey building. More than 18 000 people lost their lives.

Earlier in 2010, the Icelandic volcano Eyjafjallajökull had sent clouds of fine volcanic ash into the stratosphere, where it had been captured by the atmosphere's northern jet stream and propelled towards the crowded skies of western Europe. The dangers of dust-ingestion by aircraft jet engines are well known, and authorities throughout Europe had had no alternative but to close down their airspace. Mercifully, no lives were lost in this emergency, but an estimated ten million airline passengers were left stranded as a result—including yours truly. There was a bright side to this inconvenience, however. Because of the large-scale postponement of air travel, Eyjafjallajökull became the first volcano in history to have a carbon-neutral impact on the Earth's atmosphere. It may even have been carbon-negative, taking out more carbon dioxide than it put in—which is really rather a remarkable achievement for a volcano.

Twenty-five kilometres east of Eyjafjallajökull is one of the largest volcanoes in Iceland, a mountain by the name of Katla, whose eruptions have generally followed those of Eyjafjallajökull. Likely to cause major flooding (due to glacial melting) and further, more extensive air traffic disruption, this unstable caldera overlaid with 500-metre-thick ice is seen as a major threat to Europe's well-being if it awakens into activity. At the time of writing, two years after Eyjafjallajökull, that has not yet happened. But in May 2011, a different Icelandic volcano made the headlines. Grimsvötn became another hard-to-pronounce household name throughout the aviation industry as, once again, Europe held its breath as it waited to see the extent of the disruption it would cause. Meanwhile, down in the southern hemisphere, air travel was thrown into chaos by yet another major eruption, this time caused by a volcano in Chile's Puyehue-Cordón Caulle chain.

In the wake of these events, some have questioned whether the Earth is undergoing a kind of cosmic shake-up that signifies even bigger catastrophes to come. Is the world somehow falling apart? Are we on track for a doomsday scenario that will bring about the end of civilisation as we know it? And should we be bunkering down? Well, actually, no. Science tells us that these processes are essential parts of the normal working mechanism of the Earth, with volcanic and seismic activity concentrated around the boundaries of the crustal plates that carry continents around on ponderous geological timescales. In Japan, New Zealand and Chile, these plates are converging, with an overlap region that pushes one plate underneath another. In Iceland, they are diverging, with the eastern and western halves of the island being pulled apart at the extraordinary rate of a few centimetres a year.

The effects of these processes are certainly real, if small in comparison to the planet as a whole. Calculations show that the Tohoku earthquake, for example, moved Japan eastwards by 2.4 metres, and this redistribution of material in the Earth's crust had a marginal effect on the planet's rotation, speeding it up very slightly. The day was shortened by 1.8 millionths of a second, and the Earth's rotation axis moved laterally by about 17 centimetres, towards a longitude of 133 degrees east of Greenwich. These figures are comparable with the effects of other major earthquakes but are still below the current limits of measurement. More significantly, they are also far less than the day-to-day effects of winds in the atmosphere and ocean currents—not to mention the sloshing around of molten material in the Earth's core.

Incidentally, there is one phenomenon close to the hearts of the doomsayers that does not, in fact, occur as a result of tectonic activity. That is a change in the tilt of the Earth's axis. Confused, perhaps, by shifts in the inclination and polarity of the Earth's magnetic axis (which is known to have reversed periodically over geological time spans), some would-be pundits have speculated that the Earth's rotation axis could change direction, with catastrophic effects on climate and the seasons. But this only happens as a result of the influence of external cosmic bodies such as the Sun, Moon or planets, and it takes place very slowly.

What does Earth's tectonic activity have to do with astronomy? Quite a lot. First, it reminds us that we live on a planet that is active and alive in a geological sense. That contrasts with a world like Mars, on which, as we shall see in Chapter 9, tectonic processes have largely shut down. In comparison with the Earth as a whole, the biosphere, while vitally important to us as a species, is an insignificant and utterly powerless component. In some ways, it is surprising that the titanic forces constantly shaping the planet and its surface have not simply obliterated its inhabitants. Secondly, tectonic activity actually enhances the survival prospects of a species like *Homo sapiens* by bringing mineral and other resources to the surface, where they come within reach. Thus, it underlines the unity of all humans as inhabitants of a global village, coexisting with the cataclysmic processes that surround them. Finally, tectonic activity provides another destination for astronomy tourists. Not, of course, to gawp at the misfortunes of others in the wake of natural disasters, but to see at first hand the places on Earth where monumental planetary forces interact directly with our human world. This is hands-on astronomy of a different kind, planetary science in the raw, in which, for example, the sights, sounds and smells of a volcano on the edge of one of the Earth's great tectonic plates can be experienced with a view to gaining an ever-greater understanding of our dynamic planet—and ourselves.

I wonder if you heard the sad news that on 30 April 2009 Venetia Phair died, at the grand old age of 90. Despite her strikingly beautiful name, I can imagine you may be hard pressed to remember who Mrs Phair was. But she was famous—especially in the last few years of her life—as the only woman ever to have named a planet.

Back in March 1930, when eleven-year-old Venetia lived in Oxford and was still Venetia Burney, she heard from her grandfather that a planet had been newly discovered by an astronomer in the far-off United States and that they were trying to think of a name for it. Young Venetia was not only rather cluey about astronomy but also hooked on Greek and Roman mythology. She suggested that the Roman god of the Underworld, Pluto, might be an appropriate *alter ego* for the planet.

Such a suggestion made by most eleven-year-olds would go no further, but Venetia's grandfather happened to be friendly with Herbert Hall Turner, the professor of astronomy at Oxford University. Turner thought the idea was a cracker, although, being an Oxford professor, he probably didn't put it quite like that. Fired with enthusiasm, he telegraphed his colleagues at the Lowell Observatory in Flagstaff, Arizona, from where the planet had been discovered, and they agreed it was a splendid suggestion. Thus, on 1 May 1930, the name Pluto was formally adopted for the newly discovered planet, and Venetia was naturally rather pleased with herself. Throughout her long life, she was at pains to point out that Walt Disney's cartoon dog was named after her planet rather than the other way around.

It seems that naming celestial objects was something of a family tradition. A little more than half a century before the Pluto episode, Venetia's Great-Uncle Henry had suggested the names Phobos and Deimos (fear and dread) for the two tiny moons of Mars, which had just been discovered by scientists in the United States. They, too, were duly adopted in the world of astronomy, but sadly Great-Uncle Henry was long dead by the time Venetia's turn came around.

Why was it only in the last few years of her life that Venetia's contribution to modern astronomy became well known? The answer lies in the controversy that has surrounded Pluto in recent years as the true nature of this remote Solar System object has become the subject of intense debate. Indeed, as the discussion has descended into increasing acrimony and farce, Venetia's gentle composure has been depressingly missed.

WHEN IS A PLANET A STAR?

It was back in August 2006 that the celestial cat was set among the world's pigeons, at a General Assembly of the International Astronomical Union (IAU) in Prague. You may not have encountered the IAU before, but this venerable organisation is the governing body of world astronomy. Of necessity, it sits at the tedious end of the excitement spectrum, since someone has to dot the i's and cross the t's of all the astonishing new discoveries made by the world's astronomers and space scientists. That sort of thing isn't everyone's cup of tea, but it's an important function, and it includes responsibility for providing definitions of the various celestial objects as well as the exclusive right to name them. (Which, incidentally, gives the lie to star registry companies offering to name a star for you or your loved one on receipt of your hard-earned cash. Since they're not actually entitled to do it, it's far better to spend your money on a scintillating astronomy book like this one.)

Anyway, every three years the IAU has a General Assembly, in which matters of great cosmic weight are debated. Its 40 commissions and 76 working groups tackle such esoteric matters as extrasolar planets, high-energy astrophysics, and the origin of the Universe. Most of it is completely unintelligible to the outsider, but it's terribly exciting if you're part of the action. And part of the action in August 2006 was sorting out one particularly burning question: 'What, exactly, is a planet?'

Yes, I know. You'd think that after 400 years of looking at the sky through telescopes, astronomers would have worked out what planets are. You may also wonder whether it actually matters how the word is defined. But there's real mystery attached to this question, and it's nowhere near as daft as it sounds. It comes about because astronomy, like all the sciences, is constantly evolving as new discoveries challenge established ideas. It's just one of the things that makes science exciting—you never know quite what's around the corner.

When I was a lad, everyone knew what a planet was. It was an object orbiting the Sun and shining by reflected sunlight rather than by its own luminosity. Stars like the Sun radiate heat and light because of energy-producing nuclear reactions in their interiors, whereas planets were thought to have no energy sources of their own. There were nine planets, and their order could be recalled using various loopy mnemonics like 'My Venomously Eccentric Mother Just Served Up Napalm Pudding'. The biggest of them was Jupiter, and then, rather smaller but bedecked with a beautiful set of rings, was Saturn. Two more large planets, Uranus and Neptune, completed the quartet known eloquently as 'gas giants'—since they are big and are made mostly of gas. The rest of the planets were smallish rocky worlds, and the whole thing was very neat and tidy. Except, that is, for occasional comets and a decidedly untidy place called the Asteroid Belt, where myriads of mountain-sized boulders rolled around between the orbits of Mars and Jupiter.

I think it's fair to say that there was no single startling discovery that served to change this orderly picture. Rather, a gradual sequence of events arose from subtle new findings about both the Solar System and the planetary families of other stars in our celestial neighbourhood. To begin with, measurements made by the *Pioneer 10* and *Pioneer 11* spacecraft in the early 1970s showed that Jupiter actually gives off 70 per cent more energy than it receives from the Sun when measurements are extended beyond the red end of the rainbow spectrum of visible light, and into the invisible zone of infrared (redder-than-red) light—which is heat radiation. This put into question the idea that planets only shine by reflected light and raised the notion that a giant gas-ball planet like Jupiter might in some ways be viewed as a failed star—one that hadn't become big enough to switch on the nuclear reactions required for true stardom. And this led to a further blurring of the definitions.

In a genuine star, like our Sun, the central temperature is sustained by continuous H-bomb-like reactions in which hydrogen—the raw material of all stars—is turned into helium. The process is called nuclear fusion, and involves small atomic nuclei (of hydrogen) sticking together to make bigger ones (of helium), and releasing prodigious amounts of energy as a by-product. That's where the Sun's heat and light come from. The reactions are kicked off at the star's birth by a cloud of hydrogen being compressed, as it is pulled into a ball shape by its own gravity, and therefore getting hotter, just as the air in a bicycle pump becomes warmer as it is compressed. But if the hydrogen cloud is less massive than it needs to be to form a true star, then the resulting object is something different.

Incidentally, when astronomers discuss the masses of celestial objects, they usually lump things into convenient bundles rather than measuring everything in tonnes. There's really no future in talking about the masses of planets, stars or galaxies in tonnes—you end up with far too many zeroes. For example, the Sun's mass is about 2×10^{27} tonnes in mathematical notation—or 2 followed by 27 zeroes for the rest of us. Which is pretty meaningless, really. It's far more convenient (and meaningful) to think about the masses of celestial objects like planets and stars in relation to the mass of Jupiter. The Sun, for example, is roughly 1000 times the mass of Jupiter. 'Jupiter-masses' are therefore the units of choice for streetwise astronomers.

Back to the plot, then. Supposing that instead of being 1000 times the mass of Jupiter, like the star that became our Sun, a much smaller baby star—say, under 75 Jupiter-masses—began to form. What would happen? The smaller size would reduce the gravitational compression of the collapsing gas cloud, and its temperature wouldn't become high enough for hydrogen burning to begin. Thus, no star? Well, not quite. It turns out that if the baby star were between about 13 and 75 Jupiter-masses, a less energetic type of nuclear reaction would kick in, involving something called 'deuterium', or 'heavy hydrogen'. The resulting star would shine, but only dimly, and in infrared rather than visible light. It would become something to which astronomers have given the particularly uninspiring name of 'brown dwarf'. It's not terribly PC, but that's the way it is.

The existence of these brown dwarf stars was suspected by astronomers from about the mid-1970s, but it was not until 1995 that the first example was verified. Literally hundreds of them are now known throughout the Sun's neighbourhood in space, each typically containing around 40 Jupiter-masses of hydrogen. They represent a curious halfway stage between gas giant planets and true stars. This is further highlighted by recent studies showing that brown dwarfs actually experience weather in their atmospheres, something more usually associated with planets than with stars. And, although I was about to joke that because of their higher temperatures brown dwarfs would never have rainy weather, even this isn't true. Recent research has proved that in the atmospheres of some brown dwarfs raindrops do fall. They're not raindrops of anything as boring as water, however, but raindrops of liquid iron . . . Now that would be a downpour to avoid.

As if brown dwarfs weren't enough to blur the distinction between planets and stars, the existence of some even more bizarre dimly shining objects in the depths of space has served to increase the confusion still further. A handful of candidates for these types of exotica have been found, mostly in the star-nurseries of Orion and Taurus (where the constellation names simply signpost the areas of sky in question). The objects are popularly known as 'rogue planets', 'orphan planets', 'homeless planets' or—wait for it—FFLOPs, an absurdly appropriate acronym for 'free-floating planetary-mass objects'. I don't really like any of these names, and I'm not much keener on the one suggested by the IAU itself—sub-brown dwarfs. At least that one does give a hint as to what they are, however. They are objects containing even less material than brown dwarf stars (in other words, less than thirteen times the mass of Jupiter). They are thought to have originated in the collapse of really cute little gas clouds alongside the bigger ones in which their starry siblings—that is, brown dwarfs and normal stars—originated. (There's an alternative theory, which is, in fact, supported by the most recent research: perhaps these FFLOPs formed within planetary systems like our Solar System but were then ejected from their birthplace by some disturbance, such as the gravity of a passing star or an aggressive encounter with a fellow planet. I guess that's the origin of the name 'orphan', but perhaps 'banished' might then be a more accurate description.)

No matter what they're called or how they originated, the question that I'm sure comes to your mind is 'How do these little wanderers shine?' They don't have parent stars whose gentle light they can reflect out to the Universe at large, like the planets of the Solar System reflect sunlight. Nor are they big enough for star-like nuclear reactions to stir them into luminosity. The thinking among the pundits is that their hydrogen atmospheres may be dense enough to insulate them from the cold of space, allowing them to glow dimly in infrared light by virtue of the same sort of energy that keeps the Earth's core warm. That energy comes from the decay of radioisotopes like uranium, as distinct from the fusion processes that power stars and brown dwarfs—that is, big atomic nuclei falling to pieces rather than smaller ones sticking together. The glowing of these lonely objects by nothing more than geothermal energy sounds a bit unlikely, but calculations show that, yes, it could happen. What is far more certain, though, is that in coming years we will find more of them, as the astronomers' arsenal of space-based infrared telescopes increases in size and capability. So—if you'll excuse the pun—watch this space.

I'm sure that, by now, you're getting the picture that things are far from straightforward in the contest of planets *vs* stars. And we haven't even begun to discuss the weird and wonderful array of planets that are known to orbit other stars in the Sun's neighbourhood. These extrasolar planets, or exoplanets, have been discovered in steadily increasing numbers since the first was found, in 1995. Well over 800 were known by 2012, and the number is continuing to grow. But it was because of these exoplanets in all their wondrous diversity, together with the poor homeless FFLOPs, that the issue of exactly what constitutes a planet started to emerge as a pressing matter for serious debate.

So how did the IAU, as the body in which all power of definition is vested, deal with this? It did what any organisation would do, of course. It formed a committee. At a meeting in 1999, the IAU asked some of the leading lights on exoplanets to form a working group, part of whose brief would be to write a definition of a planet. Thus, the Working Group on Extrasolar Planets was formed. Throughout its six-year lifetime, the group was chaired by Alan P. Boss of the Carnegie Institution, in Washington DC, and it included several people whom I number among my friends. They did a terrific job, and I hope they will forgive me if I appear unsympathetic to the working definition of a planet that they eventually came up with—especially since they were at pains to point out that it was a compromise and 'did not fully satisfy anyone on the WGESP'. But, to be honest, it was rubbish.

The definition was a valiant attempt to highlight the differences between brown dwarfs, planets and FFLOPs. There's not much point in recounting all the gobbledygook here, except to quote the third and final clause, which reads: 'Free-floating objects in young star clusters with masses below the limiting mass for thermonuclear fusion of deuterium [13 Jupiter-masses] are not "planets", but are "sub-brown dwarfs" (or whatever name is most appropriate).' So there you have it. After struggling to be ultra-precise in the definition of exactly what constitutes a planet, the working group wound up by shooting itself in the foot. For the FFLOPs, at least, you can choose whatever name you think is appropriate. You might as well decide to call them 'bananas'.

AND WHEN IS A PLANET A TRANS-NEPTUNIAN OBJECT?

The difficulties encountered in working things out at the upper end of the planetary size range were eclipsed—in the public's view, at least—by what was happening at the other end. Among the diametrically challenged members of the Sun's increasingly unruly family, all was not well.

The rot had started in the 1950s, when two astronomers, Kenneth Essex Edgeworth and Gerard Peter Kuiper, independently suggested that, out in the frozen reaches beyond the orbit of Neptune, there must be a ring of debris left over from the formation of the Solar System. They argued that asteroid-sized objects in this region of space would have been too far apart to combine into larger objects and form planets, as had happened in the inner Solar System soon after its formation, 4.6 billion years ago. At best, there might be some half-finished planets out there that were too small to be seen with the telescopes of the time.

It was also thought—correctly, as it turned out—that these trans-Neptunian objects, as they came to be known, would have a high proportion of ice in their structure. They would have been perpetually too far from the Sun's heat for it to melt—or, as happens in the vacuum of space, to evaporate directly into water vapour. Eventually, in August 1992, the first trans-Neptunian object was discovered, a distant speck of light that's probably an object the size of a mountain range, taking a leisurely 289 years to orbit the Sun. It was given—and still has—the incredibly boring name of 1992 QB1, which arises from the IAU's rather inelegant naming protocol for newly discovered Solar System objects. It was also the 15 760th object discovered, in the ongoing tally of minor Solar System bodies, since asteroid number one—called Ceres—was found by one Giuseppe Piazzi, on 1 January 1801.

Significantly, Ceres itself was controversial in its day, since Piazzi and his chums at first thought it was a planet. It was found in what had been suspected to be a vacant lot between the orbits of Mars and Jupiter. But when it was discovered to be much smaller than the other planets and circulating with a clutch of even smaller objects in the same part of the Solar System, astronomers realised they had some quite un-planetary items on their hands. What could they be? It was the British astronomer William Herschel who eventually coined the term 'asteroid' to give them all a decent—if slightly comical—identity.

But back to the outer Solar System. Today, there are more than 1000 known trans-Neptunian objects. So many are known, in fact, that they can be categorised into different types, depending on the sizes, shapes and tilts of their orbits. Astronomers love to classify stuff, and this particular stuff is eminently classifiable. A minor problem is that specialists in the esoteric field of 'classifying stuff at the edge of the Solar System' don't entirely agree on what the categories should be. Broadly, though, there's an understanding that the trans-Neptunian objects divide into two types. Nearer to the Sun are objects belonging to the Kuiper Belt (or, more correctly, the Kuiper-Edgeworth Belt), which occupies a zone extending for about 2 billion kilometres beyond the orbit of Neptune. Since Neptune itself is, on average, 4.5 billion kilometres from the Sun (about 30 times the Sun–Earth distance), this is very remote stuff indeed. But trans-Neptunian objects in the second category are even further out. They are known, rather obscurely, as 'scattered disc objects', a reference to their wildly disparate orbits, and they extend out to staggering distances from the Sun—12 billion kilometres and beyond. In comparison with the scattered disc objects, Kuiper Belt objects have better behaved orbits, although they also divide into a couple of subcategories that needn't concern us here. Suffice it to say that Pluto sits in the middle of the Kuiper Belt, at an average distance of 6 billion kilometres from the Sun.

It was some unexplained irregularities in the orbit of Uranus that had prompted the search for a ninth planet, beyond the orbit of Neptune, during the early years of the twentieth century. The irregularities also suggested where in the sky astronomers should search. When, in January 1930, Pluto had been discovered in more or less the right place, there had been jubilation. At last, the ledger of gravitational forces would be balanced. So Pluto was originally thought to be a large planet—probably larger than Earth. But, as time went by and measurements of its diameter became ever more accurate, estimates of its size decreased. In a paper published in 1980, two US scientists derived a mathematical formula for Pluto's apparently diminishing girth, whimsically suggesting that by 1984 the planet would have disappeared altogether. Very droll. And so, as with Ceres a century or so earlier, jubilation over the planet's discovery eventually turned into consternation. Pluto was far too small to have any appreciable effect on a gas giant planet like Uranus. Indeed, it is now known to be only two-thirds the size of our own Moon.

The disappointment over Pluto's half-pint dimensions spurred a renewed search for a hypothetical planet that was the scapegoat for the outer Solar System's lack of equilibrium. But some people doubted there was any need for such an object. At last, in the 1980s, the idea of Planet X ran out of steam completely, when the mass of Neptune was carefully measured from the trajectory of a robotic interplanetary spacecraft called *Voyager 2*. That re-evaluation brought everything back into balance and eliminated the need for any new planets. Surprise, surprise: the discovery of Pluto had been nothing more than a happy accident.

In the meantime, however, we had a planet of minuscule proportions on our hands that seemed entirely at odds with the lumbering giants of the outer Solar System. Moreover, Pluto was in an orbit quite different from the other planets—very elongated, with a 17-degree tilt to the rest of the Sun's family. With the discovery of the first Kuiper Belt object, 1992 QB1, suspicions began to arise that maybe Pluto wasn't what it had at first seemed. Perhaps it wasn't a planet at all, but one of these pesky new icy asteroids.

Gradually, as more trans-Neptunian objects were discovered during the 1990s, a few brave souls began openly speculating that Pluto belonged among them. Most notable was Neil deGrasse Tyson, an astrophysicist and director of the Hayden Planetarium in New York City. When Pluto failed to appear among the planetarium's refurbished display of planets in February 2000, Tyson suddenly started to receive hate mail from US school kids. It was a sign of things to come.

Then, early in 2005, the unthinkable happened. A group of astronomers based at the Palomar Observatory in California announced that they had discovered a remote scattered disc object way beyond Pluto that was probably bigger than Pluto itself. Identified from sky images taken late in 2003, it had been given the provisional name of 2003 UB313. But the discovery team, led by Michael E. Brown of the California Institute of Technology, had its own pet name, borrowed from the heroine of a TV fantasy series—*Xena: Warrior Princess*. (Planet X subtly cropping up again, you see.) But there was the rub. With the discovery of Xena, did the Solar System have a tenth planet? Or were both Xena and Pluto something else, in which case there would only be eight planets?

Once again, the spotlight fell on the IAU to resolve the issue. But, as we have seen, its Working Group on Extrasolar Planets hadn't exactly covered itself in glory in arriving at a workable definition, and this was a much more acute problem. So the IAU brought together a different group of distinguished scientists, historians and science communicators to form a planet definition committee and nut out the answer. They were tasked with reporting the result of their deliberations to the IAU's General Assembly in Prague in August 2006 and duly set about their work. When they revealed their conclusions, a week before the General Assembly's decision, they could hardly have guessed how much the IAU's membership would disagree with them.

PLUTO'S PUZZLES

Ever since Pluto's discovery, back in 1930, by a young Illinois-born astronomer called Clyde William Tombaugh, people had wondered how this remote world was formed and what it might be like.

As Pluto traverses the frozen outer reaches of the Solar System, its surface temperature ranges between about –240 Celsius and a balmy –220 Celsius. Its orbital speed of less than 5 kilometres per second (compared with Earth's 30 kilometres per second) places it at the lethargic end of the Sun's family. But there are signs that past interactions between Pluto and its neighbours in the Kuiper Belt may have been more violent.

Pluto has been known since 1978 to have a large moon, Charon (usually pronounced 'Care-on' rather than, well, 'Sharyn'). Its diameter of 1209 kilometres has been measured with an accuracy of a couple of kilometres by observing its passage in front of a distant star—an event called an 'occultation'. Pluto and Charon are sometimes thought of as a binary system, because their relative sizes are fairly close—Pluto's diameter is only twice Charon's diameter. As a result, their combined centre of gravity, or barycentre, lies in the space between them rather than within the body of Pluto. This contrasts strongly with the situation for most planets and their moons, and may provide a clue to the origin of Charon. We know that in the Solar System's turbulent youth, collisions between young planets and the rocky debris left over from their construction were commonplace. For example, a collision between the baby Earth and a Mars-sized object is thought to have produced our own Moon, 4.6 billion years ago. About half a billion years later there was another bad patch, incongruously known as the 'late heavy bombardment', during which the Moon received most of the craters we see on its surface today. Is it possible that in one of these wild and woolly periods a violent collision between icy bodies in the far reaches of the Solar System could have produced Pluto and Charon? Computer simulations have shown that this is, indeed, possible, but there is at present no way of discriminating between that scenario and those in which Charon was simply captured by Pluto's gravity as it wandered past within the Kuiper Belt.

A further tantalising clue turned up late in 2005 in the shape of two more moons of Pluto—tiny objects no bigger than 150 kilometres across, now called Nix and Hydra. In July 2011, Pluto's known retinue was increased again with the discovery of a fourth, even smaller moon, as yet unnamed, while a year later, a fifth moon no more than 25 kilometres across was discovered. Nix and Hydra are known to orbit Pluto in the same plane and the same direction as Charon, suggesting that they may have formed as by-products of the same collision event. A neat and tidy theory, but only a closer look by a passing spacecraft, allowing such information as crater-number counts and surface compositions to be gathered, will provide the information needed to confirm it.

Pluto and Charon are locked in what is known as 'synchronous rotation', meaning that the two bodies always keep the same faces turned to one another as Charon trundles around Pluto in its 6.4-day orbit. The mechanism by which this has arisen is exactly what keeps the same face of the Moon turned towards the Earth—friction caused by tides raised on the two bodies by each other. No, you don't need oceans to have tides—they can occur in solid rock, and the forces involved exert a strong braking effect on the rotation of the two objects. Someday, in a few billion years' time, the Earth will always keep the same face turned towards the Moon—no doubt to the chagrin of the folk who live on Earth's Moon-less side.

The presence of a moon in orbit around a planet or asteroid has an important consequence for astronomers: it allows both objects to be weighed. And, remember Xena, that distant object whose discoverers thought was probably bigger than Pluto? In September 2005, Xena turned out to have a moon too, found using one of the two giant Keck telescopes in Hawaii. Today, Xena is no longer Xena but has been officially renamed Eris, after the Greek goddess of strife and discord—which hints at the climate in planetary science at the time. Its moon has a similarly appropriate name, Dysnomia (lawlessness)—in

Greek mythology, the daughter of Eris. Observations of Eris and Dysnomia have recently confirmed that Eris is 27 per cent more massive than Pluto, though of a similar diameter. (Therefore, it must be more dense, perhaps containing a smaller proportion of ice than Pluto.)

Apart from the obvious issue concerning their planetary status, why have Pluto and Eris become such celebrities in the astronomy of the early 21st century? The answer lies in what they might tell us about the formation of the Solar System and perhaps even about the origins of life on Earth.

If the typical trans-Neptunian object is a remnant of the original disc of debris that surrounded the infant Sun, then its chemistry would be nothing less than the Rosetta Stone of our corner of the Universe, with pristine dust grains that have been forever cold and frozen organic (carbon-containing) material that might carry the progenitors of living cells.

We already know that as well as being classified by their differing orbital characteristics, trans-Neptunian objects can be sorted in a different way into at least two garden varieties, some having a neutral-grey colouring and others, like a very distant one by the name of Sedna, being decidedly red. This may indicate subtly different cosmic histories throughout the age of the Solar System, with the reddish ones perhaps having a surface layer that has been modified by long-term effects such as bombardment by the subatomic particles known as cosmic rays. But whatever the reason for their different colours, any one of these objects that strayed close enough to the Sun would quickly develop features characteristic of a comet—a coma, or halo, formed by the evaporation of icy materials and the release of dust, and a prominent tail. There is a recognised class of exactly these types of objects in unstable orbits that may eventually fall into the inner Solar System as short-period comets; they are known as Centaurs. Half-man, half-beast. Half-Kuiper Belt object, half-comet. Who says astronomers have no soul?

The importance of this to the history of the Earth is that impacting comets are thought to have been a significant source of icy materials, such as water ice, methane and ammonia, for the planet. It is highly likely that more complex organic molecules were included in the same package, and a handful of scientists think that life itself may even have arrived in this way. Hence the extraordinary interest in investigating the various types of ice contained in comets and objects in the distant Kuiper Belt and beyond.

Larger trans-Neptunian objects, like Pluto and Eris, may have a different story to tell. With these, the process of planet formation seems to have been interrupted mid-flow, resulting in half-finished worlds that have nevertheless become big enough for their own gravity to pull denser material to the middle and, at the same time, make them spherical. This process, called 'differentiation', is likely to have given Pluto—and perhaps Charon too—a rocky core with an icy mantle. The process would have been greatly enhanced if a collision did, indeed, give rise to Charon, since the energy of the collision would have produced additional heat. Pluto's surface is known to consist of frozen nitrogen, with methane, carbon dioxide and ethane also present. However, the bulk of Pluto's icy mantle is likely to consist of water ice buried beneath the more volatile surface ices by the same process of differentiation. Its thin atmosphere, whose existence was confirmed during an occultation in 1988, is probably mostly gaseous nitrogen.

Why do we think Pluto and Eris may be half-finished planets? The evidence comes mainly from computer simulations of planet formation carried out at such institutions as the Southwest Research Institute, in Boulder, Colorado. They demonstrate that Earth-sized objects could, indeed, have formed in the outer regions of the Solar System. Why the process stopped is a mystery. But if Pluto really is a half-built planet, a close look at it would give us a unique opportunity to see planet formation in freeze-frame, providing real insights into the process.

What was obviously needed was a robotic space mission to Pluto. But there's a catch—and it's not just the extreme distance involved. Pluto's elongated orbit means the energy it receives from the Sun falls by a factor of three as it moves from perihelion (its closest point to the Sun) to aphelion (its furthest point) in its 248-year orbit. Perihelion occurred in September 1989, so by early in the 21st century the planet was already well on its way towards the zone in which its thin atmosphere will simply freeze onto its surface. And there was already evidence of seasonal changes in Hubble Space Telescope observations of Pluto's surface markings. The sooner we could get to Pluto, the more informative and interesting it would be.

TOWARDS A NEW HORIZON

On 19 January 2006, a long-cherished dream came true. A Pluto-bound robotic spacecraft called *New Horizons* was successfully fired from Cape Canaveral in Florida, using the Lamborghini of launch vehicles—an Atlas V rocket with some serious go-faster accessories. If we were going to start travelling to Pluto in 2006 on a timescale that would give us the best chance of investigating its atmosphere, we needed to get there as quickly as possible—and *New Horizons* broke all the records, leaving Earth at the highest launch speed ever achieved. It crossed the Moon's orbit in only nine hours and whizzed by Jupiter after little more than a year, the close encounter with the giant planet increasing its speed to a remarkable 23 kilometres per second. After years of planning—and a few false starts—humankind was at last on its way to Pluto.

The reliability of orbital mechanics means we can predict with pinpoint accuracy when *New Horizons* will reach Pluto. It will fly by the frozen world at 11.47 Greenwich Mean Time on 14 July 2015, passing Charon fourteen minutes later. Because of its speed (the close approach will take place at nearly 14 kilometres per second) there is no possibility of *New Horizons* being diverted into orbit around Pluto, so the 0.5-tonne spacecraft bristles with sensors to take full advantage of its brief encounter. They include spectrometers to analyse the barcode of information locked up in Pluto's rainbow spectrum, subatomic particle detectors, a long-range camera and an instrument named the Venetia Burney Student Dust Counter, which will provide valuable information on the levels of interplanetary dust in the outer Solar System. The fly-by should allow detailed mapping of Pluto and Charon as well as collection of telltale data on their surface and atmospheric composition. Alongside all the high-tech robotic sensing equipment, *New Horizons* also carries a poignant reminder of its place in human history. On board is a container carrying 28 grams of the ashes of Pluto's discoverer, Clyde Tombaugh, who died in 1997.

It's hard to overstate the importance of *New Horizons*, since our first-hand knowledge of Pluto and its environment is so sparse. The results could be the most surprising of any deep-space mission yet, notwithstanding the extraordinary discoveries about Saturn and its moons that have been made by the highly successful *Cassini* mission since it reached the planet in July 2004. But with *New Horizons* going on to target selected Kuiper Belt objects after its Pluto fly-by—and then escaping from the Solar System altogether—the excitement of new discoveries may continue well into the century.

Controversially, *New Horizons* carries 10.9 kilograms of radioactive plutonium dioxide to provide thermoelectric power for its onboard instruments. There is little alternative to this, given that the intensity of sunlight at Pluto's distance is only one 1000th of that which we receive on Earth, rendering solar panels useless. But you won't be surprised to hear that this was not the only controversy surrounding the Pluto mission.

DWARFED BY CONTROVERSY

Just seven months after *New Horizons* was launched, the IAU held its much-vaunted General Assembly in Prague. Its latest planet definition committee had agreed on a draft specification of what constitutes a planet, and this was receiving substantial press coverage. To be a planet, the committee suggested, a celestial object needed to be in orbit around a star and large enough for its own gravity to pull it into a spherical shape (a condition technically known as 'hydrostatic equilibrium'). This definition significantly extended the inventory of planets in the Solar System, since both Eris and the largest asteroid, Ceres, qualified. Moreover, there would be a significant likelihood of more to come as the exploration of the Solar System's twilight zone revealed further Eris-like objects.

Shut away from the glare of the waiting media, the membership of the IAU met on the General Assembly's final day to vote on the recommendation. But an intense debate yielded a revised definition subtly different from the committee's in that it included an additional criterion. To be a planet, went the revised version, a celestial object also had to be the dominant object in its neighbourhood, having cleared away smaller debris either by absorbing it (as the Earth does with thousands of tonnes of meteoritic dust per year) or by ejecting it through gravitational forces. Any object that hadn't done this, despite meeting the spherical shape criterion, would be termed a 'dwarf planet' rather than a 'planet'. What that meant in practice was that an object in the main Asteroid Belt, between Mars and Jupiter, or in the Kuiper Belt could not be a planet.

When the vote on this revised definition took place it was passed with an overwhelming majority. It was adopted by the IAU as Resolution 5A of the Prague General Assembly. Thus, on 24 August 2006, the world was given, for the first time, a formal definition of a planet—and it did not include Pluto. The former ninth planet, along with Ceres, Eris and two other Kuiper Belt objects, Haumea and Makemake, had become a dwarf planet.

If there had been any doubt about the public's interest in what constitutes a planet it was quickly dispelled by the outcry that followed. The headline I liked best appeared in a Newcastle (New South Wales) article: 'Pluto dumped by the übernerds of Prague.' Not just the nerds, mark you, but the 'übernerds'. Similar sentiments echoed around the world, especially in the United

States, from where the former ninth planet had been discovered. There, protest marches were held in some cities.

It was a long time before the outrage subsided, and even then it didn't do so before three US states had attempted to introduce legislation to reinstate Pluto as a planet. Only one of them, Illinois, succeeded, demonstrating that if you don't like a scientific outcome you can always legislate to overturn it. Significantly, Illinois was the state in which Clyde Tombaugh was born. And on the lunatic fringe there are still conspiracy theory websites that point to the fact that the IAU's decision was taken on the final day of its General Assembly, when only one-sixth of the attending membership was still present (producing 424 votes). They cry foul, citing rogue scientists and vested interests. They also conveniently ignore the fact that at the IAU's next General Assembly, held in Rio de Janeiro in 2009, there was no change of heart on the issue.

One consistent voice of reason in the debate has been that of Alan Stern, formerly of the Southwest Research Institute. Stern is principal investigator with *New Horizons*, and you have to have some sympathy for his view. When his spacecraft was launched, it was on its way to a planet—but now it isn't. Stern has criticised the IAU's resolution, calling it 'an awful definition; it's sloppy science and it would never pass peer review'. He cites the fact that several of the Solar System's planets, including the Earth and Jupiter, have not entirely cleared their neighbourhood of debris and therefore do not strictly meet the new criteria for planethood. Perhaps in deference to this view, the IAU executive committee announced a new type of celestial object in June 2008—the 'plutoid', which is basically a dwarf planet in an orbit beyond Neptune. Currently, Eris, Pluto, Makemake and

Haumea are the only known plutoids, but it is very likely that, as observations improve, other Kuiper Belt objects will turn out to be spherical and therefore qualify. The plutoid's definition, too, has been widely criticised, but this time because the term sounds too much like an unpleasant skin complaint.

My own view is that, while I agree the IAU's definition of a planet is not perfect, it's a lot better than what we had before—which was essentially nothing. The reclassification of Pluto shows science in action, as researchers come to terms with new information and act appropriately upon it. To have done otherwise would have been to deny what nature is telling us. My only reservation about the IAU's decisions is that perhaps the designation of the plutoid is unnecessary. It adds little to the discussion but leaves the IAU open to criticisms of gratuitous tinkering around the edges.

PLUTO AND THE ASTRONOMY TOURIST

Notwithstanding that reservation, it's clear that the Pluto saga is one of the great science stories of our age. It is also a story that is ripe for exploration by intrepid astronomy travellers as they seek out their own new horizons. There are plenty of appropriate destinations to choose from, ranging from Clyde Tombaugh's birthplace, in Streator, Illinois, to the Lowell Observatory, in Flagstaff, Arizona, from where Pluto was discovered, and Oxford, where it was named.

For our Stargazer II tour of Europe, however, we had elected to go to Prague, where Pluto was stripped of its planetary status. It was not solely for that reason, of course—there is much else of astronomical interest there. As we explored that marvellous and quintessentially European city, we sought out the restaurants in which Pluto's fate had been ardently discussed back in 2006—and drank a toast to the dwarf planet's health.

Astronomy travellers in Australia have an opportunity to see another facet of the Pluto story. At Coonabarabran, in north-western New South Wales, the Siding Spring Observatory is home to the Anglo-Australian Telescope (AAT), whose 37-metre diameter dome was, in the mid-2000s, chosen to represent the Sun in a large-scale model of the Solar System. The World's Largest Virtual Solar System Drive stretches over much of north-western New South Wales, with accurate three-dimensional representations of the planets strung out along the major highways of the district at their correct relative distances from the AAT dome. Travelling from one end to the other involves several hundred kilometres of driving, with the occasional planetary encounter bringing a whole new meaning to 'Are we there, yet?' The scale of the model, one to 38 million, means that the Earth is reduced to 33 centimetres in diameter, while giant Jupiter spans an imposing 3.8 metres. To make the Virtual Solar System Drive accessible to as many travellers as possible, the outer planets run along several highways that converge on Coonabarabran, so there are multiple versions of them. Thus, there are three Saturns, and five each of the planets Uranus, Neptune and, you guessed it, Pluto.

The Warrumbungle Shire Council built the Virtual Solar System with funding from the commonwealth government's AusIndustry Tourism Development Program, and I had the privilege of serving on the project's steering committee. When news came through of Pluto's demotion, in August 2006, we were just putting the finishing touches to each of the planet displays. It was too late to exclude Pluto from the model, but our captions had to reflect its new status. However, since every changed letter on the five affected caption boards would crank up the final price tag, we had to do some very subtle fine tuning of the wording. So, if you think the descriptions of Pluto in the Virtual Solar System Drive sound a bit stilted, that's why.

I have one last traveller's tale relating to Pluto, and it takes us neatly back to where this chapter began. When Venetia Burney changed her name to Venetia Phair in 1947, it was because of her marriage to one Edward Maxwell Phair. Edward was a teacher at Epsom College, a public (that is, private) school in the south of England, where he eventually became a housemaster and head of English. A few years ago, on the other side of the world, in the northern New South Wales town of Armidale, I gave a talk on Pluto—the same talk that later exasperated the pink-robed Messenger Nine in Berlin. Afterwards, I was approached by a delightful gentleman who, it has to be admitted, was in every sense the antithesis of Messenger Nine.

'Guess what?' he said. 'When I was a youngster, I was a student at Epsom College, and Mr Phair was one of my teachers. He was a great guy and always very popular with the students—but we used to call him Foxy Phair!'

Foxy Phair . . . or should that be Phoxy Phair? Either way, it's a marvellous scrap of trivia for the Pluto archive.

3 UNLUCKY FOR SOME

Peru's ancient skywatchers

'They're eyes,' said Iván Ghezzi. And he was quite right—they were. Out of the upright slab of carved stone in front of us peered row upon row of eyes, stylised but unmistakeable. This was beginning to get just a bit spooky.

The best I could do to rationalise these strange carvings was to guess that they might be symbolic of a great culture of ancient skywatchers. After all, yesterday we had stood in front of a series of extraordinary stone towers a few kilometres from here that had only recently been identified as an ancient solar observatory. And the memory of that place—Chankillo, in Peru—haunted my mind, as it still does today. There was no doubt that the astronomical knowledge needed by Chankillo's builders to place their thirteen square towers in exactly the right positions along the barren hilltop would have taken decades—maybe even centuries—to accumulate. Ghezzi himself had nailed down the construction of the towers to the fourth century BC using carbon-14 dating, but he'd told us that the orderly array of staring eyes now in front of us was at least 1000 years older.

As if hearing my silent musings, Ghezzi pointed out another upright panel of stone in the wall, a few metres from the peering eyes. A dozen faces, all looking extremely glum, were carved, one above the other, into the surface of the slab. They all seemed to have their eyes closed. Could they be sleeping astronomers, perhaps?

Ah. No. Between the two slabs, a third stone revealed the true meaning of these intriguing symbols. It carried the unmistakeable figure of an armed warrior, proud and gleeful in victory, wearing an elaborate headdress. 'Those aren't faces,' said Ghezzi. 'They're heads, stacked up on top of one another. With their eyes gouged out.' Oh, yuk.

As we looked with new enlightenment along this 100-metre mural wall, we could now discern a horrific celebration of slaughter and dismemberment. Severed heads by the dozen trailed blood from their eyes and mouths. Arms and legs vied with stylised human backbones to shock our unaccustomed eyes. A warrior, cut in half, his entrails hanging from the stump of his torso, gazed blindly skywards. And, in a carving described with impressive clinical detachment by the doctor in our party as 'amazingly accurate', a complete digestive tract from gullet to intestine trailed down the face of another slab. Yuk, yuk, yuk.

What came to my mind most forcibly as I went from one to another of these ancient stones was the sheer abundance of carnage. To the craftsmen who sculpted them, nearly 3500 years ago, the bloody remnants of a vanquished foe must have been a common sight. It would not have done to be squeamish . . .

As you have probably guessed, Iván Ghezzi is an archaeologist. In fact, as the former archaeology director of Peru's National Institute of Culture, he's a leader in his field. It was a tremendous privilege to have him accompany our tour to Cerro Sechín, in the coastal desert region 370 kilometres north of Peru's capital, Lima.

Looking back on it now, that particular astronomical journey has a definite air of unreality. It was the first one I'd done as a would-be astronomy tour leader rather than a private individual, and it came about almost by accident. A few months before, at the end of 2006, I'd had an out-of-the-blue phone call from someone who told me she was a travel coordinator and that she was looking for an astronomer to accompany an expedition to view a total eclipse of the Sun late in 2012. Since that was more than half a decade away, there wasn't too much in my diary. It seemed a fairly safe bet to say yes.

'There's one condition, though,' I'd said.

'OK, what's that?'

'I'd like to take a party of Australians around Europe in 2008 to celebrate the four hundredth anniversary of the invention of the telescope.'

She didn't miss a beat. 'OK. It's a deal. By the way, how would you like to get some practice, with an astronomy tour of Peru?'

Well, that was a surprise—but what a great idea.

Then, quite by chance, during the preparations for the tour, there was big news from Peru on the subject of archaeoastronomy—the study of ancient sites used by skywatchers in the remote past, and the main focus of our tour. A couple of archaeologists, Iván Ghezzi and his collaborator Clive Ruggles of Leicester University in the United Kingdom, were making worldwide headlines with their announcement that the mysterious Thirteen Towers of Chankillo constituted a previously unrecognised solar observatory that was 2300 years old.

Intrigued by this discovery, I wrote to the two high-profile professors explaining that I was about to take a party of interested Australians on a study tour of the ancient astronomical sites of Peru—and that we would be starting with Chankillo. To be honest, my email was intended mainly as a courtesy. The most I expected in reply was a polite welcome, which I could proudly read out to my little band of archaeoastronomers to demonstrate that we had the imprimatur of the masters. The least I expected was what you get from your service provider when you try to port your mobile phone number to a new carrier. Nothing.

What I actually received amazed me. In fact, to borrow the word Clive Ruggles himself famously used when the significance of Chankillo dawned on him, I was gobsmacked. A friendly message from Ghezzi told me that not only was he willing to meet us when we arrived in Lima, but he would be delighted to accompany us to the site when we visited it. This was way beyond my wildest expectations, but I soon discovered that such generosity is typical of the man.

So it was that, having seen for ourselves the Thirteen Towers—of which more later—we were gathered at the nearby older temple of Sechín, listening to Ghezzi's account of the gruesome mural art stretching before us on its outer wall. He described what is known of the cultural background underlying the scene and how it relates to the astronomical significance of the Chankillo towers. Although the builders of the temple and the towers were separated by 1000 years, their main stock-in-trade seems to have been the same: the bloody carnage of ritual warfare.

SKYWATCHERS AND RITUAL

Perhaps more than most folk in today's world, astronomers are tuned in to the idea that primitive people placed great significance on the things they saw in the sky. You have only to think of our familiar join-the-dots constellations and the stories attached to them. They're based on the star patterns of the ancient Greeks, which in turn owe their origins to earlier Sumerian constellations. But many ancient peoples identified their own star patterns and evolved legends to go with them. Often, as in the case of the Indigenous people of Australia, those legends had practical purposes. For example, they defined seasons when it was worthwhile hunting for a particular food—berries or insect larvae—or, in other cultures, when it was the right time to plant crops.

In the coastal deserts of Peru, human settlement clung to the flood-plains of Andean rivers, which, even today, cut green swathes through the pale grey-brown of the desert soil. Water meant everything, and the ebb and flow of the rivers—which in reality are determined by distant snow-melt—were believed to be controlled by supernatural beings. Perhaps these were the ancestor gods who were considered by many ancient Peruvian cultures to intervene directly in everyday affairs and who were closely linked with the phenomena of the sky. This tinkering with the world by unseen gods was all well and good, but what happened when they disagreed among themselves? Well, obviously, they would have a go at one another in heavenly battles—and maybe the outcomes of the battles would determine who in the human world got water and who didn't.

But here's the twist. In the belief system of many ancient Andean people, a supernatural battle would have its parallel on earth, in the form of a ritual battle founded on religious doctrine—a holy war. Today, anthropologists distinguish between such sacred wars, with their specific religious aims, and true warfare, in which there were wider socio-political goals. The ultimate aim of true warfare was the total destruction of the enemy, while in a ritual war the loss of life might be quite limited. On the other hand, the ritual butchery of the vanquished in early religious warfare could assume truly barbaric proportions.

To initiate a ritual war, it was believed, the ancestor gods would speak to the leaders of their respective communities in dreams or through astronomical phenomena—events in the sky. Belief in such messages was fundamental to the process, and, clearly, astronomical observations could have serious consequences, initiating battles with rival communities. It's just as well that today's astronomers don't have to worry about such outcomes—or, at least, if they do, their battles are fought in learned journals and are seldom life-threatening.

It is Iván Ghezzi's thesis that many of the most spectacular archaeological sites of the Casma Valley—where both Sechín and Chankillo lie—were ceremonial settings for such ritual battles. He has good reasons for believing this.

Chankillo itself is an enormous archaeological complex, covering several square kilometres of the desert floor. While little more than foundations remain today, the buildings and plazas that once occupied the area were clearly designed on a grand scale. Set high on a hilltop at the western end of the site is its most substantial and prominent feature. The imposing construction consists of two roofless circular buildings about 40 metres in diameter, with a third, rectangular building of similar size close by. The three structures stand on an earthwork platform and are surrounded by two rough stone walls approximately triangular in plan and massively built—up to 8 metres high and 6.5 metres thick. This is major structural engineering.

It is principally from wooden lintels in the doorways piercing these walls that the accurate carbon-14 dating of the complex has been made. Carbon dating from other contemporary plant samples (such as seeds and fibres) is in good agreement with the 2300-year estimate of its age. The whole site also shows evidence of damage due to seismic activity, which has produced characteristic triangular breaks in the masonry. Peru is no stranger to earthquakes, for the same reason that Chile is no stranger to volcanic activity, as we saw in Chapter 1. Plate tectonics again.

The question that puzzles archaeologists when they examine this structure at Chankillo is what exactly it was for. It is often referred to as a fort, and it certainly looks like one, with its massive defensive walls. However, if that is the case, what was its strategic purpose? As well as being 180 metres above the valley floor, it's more than 2 kilometres from the fields and water sources that it might have been meant to defend. It has no water-storage facilities of its own, so it would be quite unable to endure a siege. And, strangely for a fort, the stonework reveals that the wooden bars that would have latched its gates firmly shut were not actually on the inside of the main walls, but on the outside. Eh? Surely that can't be right. Could it be possible that enemies wanting to gain entry would simply have been able to let themselves in without even having to knock?

To many archaeologists, all these aspects strongly suggest a ceremonial rather than a defensive purpose. Perhaps this was a seat of power, symbolising the holy place of a particular community. In that regard it has more in common with a temple than a fort—a kind of primitive cathedral. But there is other evidence, uncovered by a team of archaeologists led by Iván Ghezzi, that points to its importance in ritual conflict. Remnants of parapets have been found surmounting the inner platform on which the three buildings stand. They prove that this place was genuinely intended to be defended against an attacker, notwithstanding the latches on the outer walls being the wrong way around. Moreover, thousands of round stones of a uniform size are scattered on the hillside and plain below it. These are sling-stones, which would have been gathered from the riverbed 2 kilometres away. In the hands of warriors, they would be lethal weapons. And many fragments of ceramic figurines depicting such warriors with a variety of weapons have also been found.

There's other evidence, too. That rectangular building within the main complex aligns with the direction of sunrise at the summer solstice—the time of year when the midday Sun is at its furthest south, on 21 December. This alignment is maintained in the foundations of other buildings that stretch away several kilometres to the east. Of these, we will hear more shortly. But the uniform alignment suggests that the rectangular building on the hill was of great importance to the whole site—a focal point for the activities carried out there. Indeed, it can be seen from almost anywhere in the district.

What was in this building? It contained several rooms. Raised platforms in two of the rooms suggest that ceremonial rituals took place there—we have no idea what they were. The remains of decorated pillars in these two rooms have given the building a name—the Temple of the Pillars—for their function seems to have been primarily ceremonial rather than merely holding up the roof. In any case, the roof would have been only a lightweight fabric or wooden structure.

Remarkably, one room in particular, a kind of inner sanctum, shows evidence of having been systematically destroyed. A thick layer of dirt and stone has been piled into the room in a manner quite different from the damage caused naturally by earthquakes. Indeed, most of the stony material is of a type not found at Chankillo, so it must have been transported from elsewhere. It is as if the intention was to annihilate the room completely. Along with whatever gods and altars it had housed, it was meant to be eradicated from memory altogether. And all the signs are that the eradication was done by intruders from distant parts.

These pieces of evidence led Ghezzi to conclude that the Chankillo structure was neither a fortress nor a temple, but an amalgamation of both. During its period of occupation, it had enormous religious significance to the people of the Casma Valley, but the systematic destruction of the Temple of the Pillars marked the end of its era of dominance. And from this he draws broader conclusions:

> At least for the construction of the Chankillo fort, [the] settlements and their populations had lesser

> priority than ceremonial spaces in the assignment of public labour for defensive works; it suggests that a major goal of warfare may have been to attack the seats of religious power . . . The threat of total destruction may have been key to justifying leadership [together with] the mobilisation of public labour to erect massive fortifications to protect gods and their temples from the dangers of a world in which holy wars were fought to destroy them.

For the astronomers of Chankillo, that threat may well have been the motivation for their careful observations. And the Thirteen Towers remain today as highly suggestive evidence of this.

CHARIOTS OF THE GODS AND ALL THAT

Think of ancient civilisations in Peru and it's a fair bet that your mind will turn first to the Incas. The amazingly precise stonework of Inca buildings speaks of a culture that was highly accomplished in both the theory and the practice of architecture. Their characteristic trapezoidal doorways, windows and niches bear testimony to a deep understanding of the kind of civil engineering necessary to withstand frequent earthquakes, and it is clear that such knowledge—like that of the ancient astronomers of Chankillo—did not appear overnight.

It is surprising, then, that this extraordinary culture lasted little more than 100 years, from the expansion period of King Pachacutec, in the 1430s, to the capture of the last Inca monarch, Atahualpa, by Spanish conquistadors, in 1532. Most of the Inca structures that remain today date from this period, and the associated ceramics and textiles that are preserved in Peru's museums tell of a creative and artistic civilisation. Perhaps it was that accomplishment that goaded a few hundred power-crazed conquistadors to wipe out twelve million Incas after Atahualpa's capture and execution, in 1533. The technological prowess of the Spanish made that task as easy as pulling the trigger of a flintlock.

The principal god in Inca religion was the Sun, followed by the creator god, Wiracocha, and then the Moon and various other deities. It's therefore no surprise to find that many Inca buildings and complexes align with the direction of sunrise or sunset at the winter (21 June) or summer (21 December) solstices. On these dates, the midday Sun is at its furthest north or south in the sky respectively. As a result, the compass bearing of sunrise (or sunset) is at its most northerly (in winter) or southerly (in summer) point along the eastern (or western) horizon. I hope that makes sense—it certainly did to the Incas.

Such an alignment is found at one of the Incas' principal religious sites, the Temple of Wiracocha, at Raqchi, in southern Peru. This vast cathedral-like structure would have been awe-inspiring in its heyday, with its central roof-bearing wall 12 metres high and a total extent of 92 metres in length and 25 metres in width. Streets containing smaller buildings lay to the south, aligned with the winter solstice sunrise, and the whole is still surrounded by a wall 5 kilometres long. An artificial lagoon within the wall is fed by waterways built in stone by Inca hydraulic engineers, and one suggestion for the intended purpose of the lagoon is that it could have been used to observe a reflected image of the Sun in worship. A neat idea, but one that is nothing more than speculation.

Raqchi is about 100 kilometres south-east of Cusco, the principal city of the Incas and today the hub of archaeological and tourist activities in this area of the country. Founded in pre-Inca days, around the year 1250,

Cusco lies at the head of the Sacred Valley of the Urubamba River, which cuts through lofty mountains to the north-west of the city. Prominent among the Inca villages that line the valley floor is Ollantaytambo, site of a royal palace and associated barracks to house large numbers of soldiers.

Once again, the winter solstice sunrise played an important ceremonial role in Ollantaytambo, but this time its position was marked not by an alignment of streets but by a prominent feature on the near-vertical wall of a sacred mountain, as viewed from the terraced hillside behind the town. It is thought that the rising of the well-known Seven Sisters star group—the Pleiades—also had religious significance, and another feature on the same sacred mountain marks the place where this star group comes into view.

The jewel of the Sacred Valley—and perhaps the jewel of all Peru—is the astonishing mountain-top Inca settlement of Machu Picchu, further down the Urubamba River. Known to science only since 1911, when it was rediscovered by a gentleman named Hiram Bingham, of Yale University, in Connecticut, this extraordinary city in the clouds escaped destruction by the Spanish, perhaps because it was abandoned unfinished by its Inca builders at the time of the conquest. Now recognised as one of the New Seven Wonders of the World, Machu Picchu has become a Mecca for tourists, although archaeological research on the site continues.

Machu Picchu, too, contains structures that align with the rising Sun at the winter and summer solstices. Indeed, one D-shaped building within the complex is described as an observatory, though it certainly isn't an observatory in the modern sense of the word. A clue to its significance, however, comes from the fact that its stonework is some of the finest on the site, and it was the standard practice of Inca stonemasons to grade the finish of their work according to the importance of the structure. For me, the most striking thing about the building is the amazing precision with which the lower courses of stonework blend into the rocky outcrop on which it stands, giving the impression of a structure simply growing out of the Earth. The subtlety of those Inca architects is awe-inspiring.

The expertise that went into the design of such fantastic structures was probably gained over many centuries, and nowhere is that more apparent than at another iconic site of Peru's spectacular past—Nazca. Down in the southern half of the country, on a desert strip 60 kilometres from the Pacific Ocean, lies the Nazca Plateau. Like the splendour of Machu Picchu, its secrets are a twentieth-century rediscovery, for it was only when commercial aircraft began flying over the region, in the 1920s, that reports of strange lines and trapezoids crisscrossing the plateau began to emerge.

It is the nature of the desert surface in this part of Peru that allowed the Nazca people of the Early Intermediate Period (250–600 AD) to leave their indelible mark for posterity. The desert is covered with a layer of brownish pebbles, which can be removed to reveal the lighter-coloured soil beneath, and that is how most of the lines were made. The fragile construction has been preserved by the dry, almost windless climate of the region, allowing modern archaeologists to gain unique insights into the rituals of those early Peruvians—for we now believe that ritual is what the lines are all about.

In fact, there are thousands of lines—most of them ruler straight, and often running for tens of kilometres—traversing the plateau in many different directions between the towns of Nazca and Palpa. Moreover, the survey work of a US archaeologist, Paul Kosok, and his more famous German protégée, Maria Reiche, in the 1940s and 1950s, revealed that there are also gigantic figures of living creatures—birds, fish and land animals—scattered across the desert. This is astonishing stuff, and extremely rare, although there are a couple of other places within the Americas where similarly inscribed figures can be found. The scale of the Nazca figures beggars belief, however, with the largest ranging over almost 500 metres. While many are just a fraction of this size, the fact that the figures can only be seen properly from above has given rise to many wild speculations about how and why they were constructed. Most famous are the ideas of the Swiss author Erich von Däniken as expounded in his 1968 book *Chariots of the Gods? Unsolved Mysteries of the Past.* Von Däniken decided that the lines, together with their wider, trapezoidal counterparts, were evidence of landings by alien spacecraft and that the figures were designed by said aliens to be viewed from above. Great stuff if you're into science fiction, and it certainly proved a best seller for old Erich. But it's a fantasy story whose downside is that it wholly undervalues the achievements of the Nazca people themselves.

Modern research has demonstrated that, far from requiring alien expertise, the lines and figures, or geoglyphs, could be made easily by primitive people using the technology available to them. They would have scaled up the pictures from smaller versions using sticks and ropes in a technique reminiscent of the toy pantographs some of us had as kids. Evidence of this has been found in the form of wooden marker pegs left behind by the creators of the lines. The Nazca people were clearly expert surveyors, marking out the desert with remarkable precision, and if they were a little forgetful in leaving behind their marker pegs, it has worked to our advantage. Carbon dating of these samples has given us our accurate estimate of the age of the lines.

However, although we're confident that we understand the techniques used to create the lines and figures, we're a lot less certain about why they were made. For a time, primitive astronomical observations were favoured as the motivation for the lines' construction. It was Paul Kosok who, in 1941, found that among the lines were some pointing in the direction of the solstice sunrises and sunsets. As a result, he declared that Nazca was 'the largest astronomical calendar in the world'.

The studious Maria Reiche followed that interpretation, discovering alignments with the rising and setting points of bright stars (whose positions were corrected back in time to the Nazca period) as well as those of the Sun. But the problem with this approach is that there are so many dratted lines—and so many bright stars—that you can interpret them in almost any way you choose. Why should those that seem to align with celestial objects mean any more than those that don't? As Clive Ruggles said when he was first asked to investigate the astronomical alignments of the site at Chankillo, 'Inside I was thinking "yeah, yeah, yeah"—people are always saying this to me.' In archaeoastronomy it pays to be sceptical, and a wholly astronomical reason for the Nazca lines now seems improbable.

Perhaps the best explanation for the significance of the lines comes from the fact that many of them seem to align with small hills on the plateau. Maybe these hills are somehow representative of the larger mountains in the region (including a giant 2200-metre-high sand-dune, Cerro Blanco), which were known to be reservoirs of water. Once again, the supply of water was the key consideration in the lives of the Nazca people—and once again they were faithful custodians of that precious resource, even building underground aqueducts to protect it against evaporation. Did the Nazca lines form a gigantic map of the region, with ceremonial significance in the supply of water by the gods? And were the figures part and parcel of this—animals and birds lovingly drawn in the desert to please their heavenly masters?

Recent research on shamanistic rituals still practised today in parts of South America suggests that shamans (holy men) of the Nazca period used hallucinogenic drugs to create the illusion that they were flying over the desert, interceding with the gods on behalf of their human congregation—as I guess you do under such circumstances. Perhaps the figures were intended for the spirit eyes of the shamans, to encourage and support them during their flight?

Other intriguing research suggests that the figures may have had a processional significance, allowing worshippers to walk along their outlines in an act of gratitude to the gods for the continuing supply of water—with a strong hint that it would be really nice if the supply kept on coming. This idea seems particularly attractive, given that most of the figures are depicted by a single continuous line that would allow an out-and-back procession, without the walkers having to retrace their steps.

The mystique of the Nazca lines remains undiminished for today's visitors as they are herded into the succession of half-hour flights over the region. Crash-landings are not uncommon, but the trip is worth the risk. Seeing these extraordinary markings for yourself from the cabin of a small aircraft gives you a direct link with a culture whose ideals and aspirations were wholly different from ours—but which you can't help but respect deeply.

THIRTEEN TOWERS

Cusco, Ollantaytambo, Machu Picchu, Nazca—no archaeoastronomy tour of Peru worth its salt would miss any of these places, and ours didn't. Moreover, we threw in a few more for good measure. Unfortunately, there is no space here to describe the floating Uros islands of Lake Titicaca or the extraordinary funeral chambers of Sillustani, built over a span of 2000 years. The archaeological sites of Sacsayhuaman (memorably pronounced 'sexy woman'), in Cusco, and of Lima itself must likewise go uncelebrated here.

For most of us on the tour, those stars of archaeology were eclipsed by one epic journey, the one we made at the start of our sojourn in Peru. In our small way we made history, by becoming the first study group from Australia—and among the first in the world—to see the extraordinary Thirteen Towers of Chankillo for ourselves. And our pilgrimage to this largely unvisited area of northern Peru left an indelible mark on us all.

We were encouraged by our expert guide, Iván Ghezzi, to make it our goal to witness the Sun setting behind the towers in a re-enactment of the way it would have been viewed in the towers' heyday. That turned into a race against time, as our coach sped northwards up the Pan-American Highway from Lima to Chankillo and then struggled to negotiate a couple of dozen kilometres of impossibly narrow farm track to get within walking distance of the site. I think we probably would have made it had it not been for the attentions of the police. No, the coach driver wasn't speeding, or drunk, or driving dangerously. The cops weren't interested in any such misdemeanours, or in the roadworthiness of our vehicle—or even in the behaviour of its passengers. (Ghezzi told me later that if the coach driver had failed to stop, the police probably wouldn't have given chase: they would have been too worried about whether they had enough fuel to get back home.) All the police wanted to know, it turned out, was whether the driver's paperwork was in order. Fine—but they seemed to want to know that every 20 kilometres or so. That made the 370-kilometre trip very, very slow, and it took a lot longer than either we or our driver had expected.

When we arrived at the Thirteen Towers, the Sun was already low in the sky, and, unusually, was setting behind banks of cloud. So we didn't see any breathtaking astronomical alignments. But what we did see in the fading light left us with a spine-tingling impression of the utterly amazing place that is Chankillo.

What are the Thirteen Towers? Why are they so significant—and what is it about them that made Iván Ghezzi and Clive Ruggles household names in the world of archaeology after they announced their discoveries in March 2007?

In the desert to the east of the big hilltop temple-fortress described earlier are many archaeological remains. Foundations of gigantic rectangular structures litter the desert floor, their alignment with the summer solstice sunrise clearly visible in satellite imagery. These are the remnants of buildings and plazas, the main fabric of the ceremonial complex of Chankillo. Running northwards through the middle of the complex is a line of low hills, and the northernmost one—a prominent ridge rising tens of metres above its surroundings—is surmounted by a row of thirteen square towers. Their solid stonework construction and regular spacing (of about 5 metres from one tower to the next), together with the provision of staircases running up the north and south walls of each tower, speak of an important purpose whose identity has, until recently, remained a mystery.

The 200-metre-long line of towers runs almost exactly north–south, although there is a well-defined bend in the alignment of the three southernmost towers. They twist around to the south-west, following the line of the ridge. From points to the east and west of the towers, observers are presented with the extraordinary spectacle of a skyline punctuated by a series of regularly spaced notches formed by the gaps between the towers. This suggested to Ghezzi and Ruggles that the towers had some astronomical significance, and their subsequent research has borne this out.

The scientists identified two significant points among the low-lying ruins of the complex from where observations of the rising and setting Sun could be made. On the western side of the towers (from where sunrises would be observed) is the open end of a long corridor whose walls would originally have been more than 2 metres high. A doorway at the end of the corridor would have faced the line of towers, about 230 metres away. On the eastern side (corresponding to sunset observations) are the remains of a small building 6 metres square, which also seems to have had a doorway facing the towers. These two doorways define an east–west line that bisects the line of towers. Ghezzi and Ruggles realised that, seen from the locations of the doorways, the row of towers corresponds in length to the range of compass bearings that the rising or setting Sun would move through during the year when calculated for 300 BC. Thus, at any time of the year, the rising or setting Sun could be observed and its alignment measured relative to the series of artificial notches along the hilltop.

Effectively, then, the builders of the Thirteen Towers had graduated, or calibrated, their skyline. By noting the position of sunrise or sunset they could estimate the time of the year—the date—accurate to just a few days. This contrasts strongly with other Peruvian sites in which the sunrise and sunset positions only at the solstices themselves (the extremes of the rising or setting Sun's daily motion along the horizon) are marked. The conjectured purpose of the line of towers is therefore that it served as a giant calendar, allowing the high priests of the community to dictate with certainty when crops should be planted or when religious ceremonies should be enacted—or, perhaps more importantly, when one of those barbaric ritual wars should be kicked off.

Iván Ghezzi has gone further in his speculations about the site. He has pointed out that the structure of each observing point, on the east and west sides of the towers, is quite different from that of the other, the western one being very restricted in its accessibility and the eastern one being in an open space overlooking a large flat area. This suggests that perhaps observations of sunrise were made privately by a small number of high-ranking officials crowding into the narrow outer corridor of a sacred building. Sunset observations, on the other hand, were perhaps made out in the open and could have had an audience of thousands. This might have been where leaders told the troops about the next earthly onslaught in the eternal supernatural battle—with supporting evidence provided by the setting Sun right behind them.

Perhaps, more broadly, the towers were about power and control within the society that built them. Ghezzi and Ruggles speculate that those who controlled the buildings would have seemed to everyone else to have control of the Sun itself, inasmuch as they had knowledge of the Sun's gradual motion along their calibrated skyline. Thus, the Thirteen Towers made a strong political and ideological statement to the uninitiated masses.

But what does Chankillo tell us about astronomical awareness in the rather incongruously named late Early Horizon period, around 300 BC, when the towers were built? Were the astronomer-priests responsible for its construction interested only in crops, holy days, warfare and power? Or was their understanding of celestial events more complex? At the very least, they recognised the most elementary aspect of cosmic mechanics—namely, that the Sun appears to move northwards and southwards in the sky following an annual cycle. Today, we know that this is due to nothing more than the tilt of the Earth's axis to the 'vertical' defined by the planet's orbit. This tilt is only about 23.5 degrees, but it produces the dramatic changes of seasons experienced in temperate latitudes. As the Earth moves around the Sun, the planet's tilted axis points in a fixed direction in space, meaning that the midday Sun is overhead alternately north and south of the Equator, gradually moving from one to the other over six months, and then back again. As Peru is close to the Equator, seasonal changes are less pronounced than in Tasmania, for example, but they are eloquently displayed by the motion of the Sun's rising and setting points along the horizon.

Personally, I believe that the astronomers of Chankillo had some more-subtle insights, because the Thirteen

Towers still hold mysteries that we can't explain. One is that their height varies systematically along the hilltop. We know from their construction that, despite significant earthquake damage, the towers are the same heights today as they were when they were built, so their varying heights were quite deliberate. But why should that be the case? Since the towers' heights increase towards the northern end of the row, they combine to effectively reduce the slope of the artificial skyline compared with the natural line of the hilltop itself, so that the tops of the towers follow more closely something called a 'meridian of hour angle'. Basically, this means that the Sun will always cross the tops of the towers at roughly the same time of day throughout the year. Or is this an over-interpretation that would make Clive Ruggles think 'yeah, yeah, yeah' once again?

Another mystery surrounds the visibility of the towers themselves. Because of the south-westerly bend in the line of towers at its southern end, only eleven of them can be seen from the eastern (sunset) observation point. The other two are either on or below the local horizon. From the western (sunrise) side, all thirteen towers are visible. This is the side that Ghezzi speculates was accessible to only a few high priests, rather than the general population. Perhaps the two most southerly towers had a special sacred significance.

But there's a further subtlety. If Ghezzi and Ruggles' identification of the two observing points is correct, the eastern point is slightly closer to the line of towers than the western point, meaning that the towers still cover the full range of compass bearings followed by the setting Sun even though two of the towers aren't visible. Perhaps this was nothing more than a natural way of compensating for the twist in the hilltop that hid the southernmost towers from view at sunset. But it's also true that having the two observation points at different distances from the line provides additional information to an observer who watches both sunrise and sunset on the same day. Potentially, it would allow the astronomical calendar to be calibrated more finely, perhaps even allowing the exact day of the year to be determined. That would be a remarkable achievement for a primitive society.

No doubt, continuing archaeological work on the Thirteen Towers will give us more insights into their construction and use. The studies of Ghezzi and Ruggles have prompted considerable debate, and at least one archaeologist has doubted the significance of the towers as a horizon marker. Ritual processions using the staircases are suggested by this researcher as their primary purpose.

Whatever the outcomes, though, there will be no diminution in the almost magical aura that surrounds the extraordinary structures. For me, the memory of my first sight of Chankillo will always remain. Approaching the line of towers nearly end-on in the weak late-afternoon sunshine with my companions, I became aware that the desert country we were walking over was littered with ceramic fragments. They were everywhere and were readily recognisable as parts of earthenware vessels. Many of them would have been broken in ritual ceremonies, and Ghezzi assured us that they date from the main occupation period of Chankillo, 2300 years ago. We could have picked up handfuls of them—though, of course, since this was an archaeological site, we didn't. 'Take only memories; leave only footprints' is the mantra. But there was no more poignant a reminder that the haunting structures in the empty desert before us were once the focus of a thriving population who lived, breathed, ate and drank—and were, in essence, just like us. As darkness fell and we hiked back to our waiting coach, we could almost hear the ghosts of those ancient skywatchers bidding us farewell.

4 STARGAZERS BEHAVING BADLY

Astronomy's controversial awakening

Marnie was a little peeved. As the coordinator of our 2008 Stargazer I tour, of the great European centres of astronomy history, she had bigger things on her mind than a missing bike, but nevertheless the irritation showed. You would think that with nearly 1000 bicycles for hire on this little island—almost three per head of the population—there would have been plenty to go around. But, on this lovely September afternoon, when a leisurely 4.5-kilometre ride from one end of the island to the other was such an inviting prospect, they were at a premium. And Marnie's had been pinched. Worse, it had been pinched by one of the participants in the tour. And, of course, the customer is always right . . .

You would be forgiven for thinking that the title of this chapter promises a titillating exploration of the misdemeanours of the Stargazer tour's participants. After all, twenty-odd lively people a long way from home can get up to all sorts of tricks. But, even allowing for the old maxim of 'What goes on tour stays on tour,' I'm afraid the mix-up with bike hire was about as spicy as it got. Rather, this chapter has a somewhat wider aim. For me, the notion of stargazers behaving badly brings to mind lots of real episodes from a lifetime spent in the trade. One of the free extras that come with a career in national observatories is that you interact with a large and representative cross-section of the professional astronomical community, as its members travel from their various learned institutions to use your facility's telescopes. It provides a unique opportunity to observe the observers, so to speak. And it soon becomes clear that astronomers, being human, are capable of all kinds of behaviour—including the bad kind.

Without mentioning any names (for many of these people are still working scientists), I could tell you about high spirits leading to irreplaceable glass photographic plates being sat on and broken. And late-night parties leading to large telescopes equipped with delicate and expensive instruments being rained on, when the weather changed. And blood-curdling expletives being hurled about when dome interior lights were switched on at the wrong time, and even worse language when sensitive instruments were switched off at the wrong time. Some astronomers have torn pages out of log-books so the next group of observers using the telescope wouldn't know where it had been pointing and therefore be able to 'steal' their target objects. Other astronomers seem to have been more interested in stealing their colleagues' ideas—or, occasionally, their spouse. And, lest I should be accused of throwing stones in glass houses, I freely admit that one of the people who dropped snowballs onto passers-by from the tower of Strasbourg Cathedral during a wintry 1980s astronomy conference was me. Disgraceful.

All sorts of considerations, mostly legal, prevent any further exploration of these particular issues—at least for a few decades. However, they really only scratch the surface. Astronomers have been exhibiting bad behaviour of one sort or another for centuries, and many significant historical events have been characterised by less-than-wholesome undercurrents. Just a few of them are revealed in this chapter.

Meanwhile, back in the 21st century, our doughty tour coordinator turned the unexpected loss of her bicycle into a highlight, relaxing on the warm grass and trying to imagine what that same spot would have been like 400 and more years earlier. The place where she lay, alongside the rest of us who weren't cycling, was the epicentre of the island's claim to fame. This place was Stjerneborg, the Castle of the Stars, an observatory built by the astronomer Tycho Brahe in 1586 on the fabled Danish island of Hven. The facts that the island is these days called Ven and since 1660 has belonged to Sweden didn't really seem to matter on that balmy autumn day. Nothing could change the lasting impression that this was one of the most enchanting places in the world.

A PRODUCT OF HIS TIME . . .

The venerable Tycho Brahe was the greatest observational astronomer of the pre-telescopic age—truly the Lord of the Stars. Known to posterity by the Latinised version of his first name (Tyge), adopted while he was a student, he lived on the eve of the telescope's emergence from obscurity in the early 1600s. Born of noble Danish parents in 1546 (and, by coincidence, sharing a birthday with your humble author), Tycho was brought up in rather unusual circumstances. In a display of behaviour that would be considered pretty bad by today's standards, his aunt and uncle stole him from his parents when he was a baby and raised him as their own son. Sixteenth-century Denmark was more tolerant of such events than we are, however, and his natural parents simply took it in their stride, shrugged their shoulders and got on with producing more kids.

Tycho is best remembered for his amazingly accurate observations of the positions of planets and stars made in the 1580s and 1590s with advanced sighting instruments of his own design. Most of these were located at his two observatories on that tiny island in the Øresund, the strait separating modern-day Sweden and Denmark. Uraniborg, the Castle of Urania, came first, completed in 1580. It also served as Tycho's home and laboratory, but nothing remains of it today save its modest outline amid the restored splendour of the lord's gardens. A few metres away is Stjerneborg, completed in 1586 and sunk into the ground to provide better stability for Tycho's observations than the rickety wooden platforms of Uraniborg. It has fared rather better, and there are still fragments of the foundations of his instruments to be seen. Stjerneborg now forms the backdrop of an audio-visual diorama for visitors, incorporating these important remnants. The exquisitely accurate measurements made in Tycho's observatories eventually led to the laws of planetary motion established by his former student Johannes Kepler, who was, without doubt, the greatest mathematician of his day; his laws are still taught as the basis of orbital dynamics.

Besides being an astronomer of formidable skill, Tycho was also the first modern-style director of a scientific institution. He was accomplished in what would today be called project management, organising his financial and human resources to maintain Renaissance Europe's foremost research centre. But he is also remembered for his arrogance and his fiery temper, and it was these aspects of his personality that got him into all sorts of trouble in the fateful December of 1566.

Tycho was a twenty-year-old university student at the time, and, in common with all respectable astronomers of his day, was also a keen exponent of astrology. While today we know that the art of divining personal insights from the stars has no basis in fact—even if some do find it irresistibly fascinating—in Tycho's era it was taken seriously. Indeed, while the word 'astronomy' is derived from the Greek for 'numbering the stars', 'astrology' comes from 'words about the stars'. In the sixteenth century those meanings were nearer to the surface than they are today.

The late Victor E. Thoren (one of the twentieth century's most respected scholars of Tycho's life and work) suggested that it was an erroneous astrological prediction that led to the unfortunate duel in which Tycho lost most of his nose. It seems the sequence of events began with Tycho's deep and meaningful deliberations on the astrological significance of a lunar eclipse on 28 October 1566, an event in which the Moon passed through the shadow of the Earth. He decided it foretold the death of Süleyman the Magnificent, the 70-year-old sultan of the Ottoman Empire. To be honest, that would have seemed like a pretty safe bet for any would-be astrologer, since living to the age of 70 was a truly remarkable achievement by the standards of the day. So Tycho broadcast his prediction widely.

Unfortunately, it soon emerged that Süleyman had died several weeks before the eclipse, and Tycho was ribbed mercilessly over the blunder. Chief among his critics was Manderup Parsberg, a fellow student and distant relative of the young nobleman. According to Thoren's interpretation, several days of intermittent bickering between these two eventually flared into something more serious on 29 December. Rather inconveniently, the matter came to a head at a dinner party in, of all places, the house of a professor of theology. No doubt the feisty Tycho reckoned that a duel would finish this acrimonious debate once and for all. In the event, however, he was the one who almost got finished.

The contest was very short and seems to have been stopped through the intervention of the other guests—understandably put out by this indecorous interruption to their dinner. (Probably, someone got a bit of nose in their soup. That would certainly be off-putting at a dinner party.) Tycho, covered in blood, may not at first have realised just how lucky he had been. The blow that had sliced open his forehead and removed the bridge of his nose had missed his eyes by millimetres. It is truly remarkable that the person who was destined to become the most accomplished naked-eye astronomer in history was spared by a whisker from losing his sight altogether.

The prosthetic noses that Tycho developed later in life are well documented. There was a copper one for weekdays and a silver-and-gold-alloy model for Sundays—and he was never without the jar of ointment he used to stick them on with. Pity help anyone who got in the way when he sneezed. But while the great astronomer carried the disfigurement of his youthful bad behaviour to his deathbed, there is evidence that he bore no grudge against Parsberg. That is surely an illuminating insight into his character.

As he grew older, Tycho matured into a great scholar, an erudite man whose company was sought by monarchs and nobles as well as by his fellow scientists. But there is one other aspect of Tycho's life that calls his behaviour into question, and again it was the consequence of his high station in life, as well as of the era in which he lived. It seems that, for all his being a generous teacher and devoted husband and father, he rode roughshod over the peasants of Hven. He was, in fact, within his rights to make significant demands on the islanders, as Hven had been gifted to him, in May 1576, by his patron, Good King Fred—Frederick II of Denmark and Norway—and a proportion of their labour was due to Tycho as their feudal lord. The trouble was, the islanders had previously considered themselves to be freeholders in their own right and strongly resented the dues they were forced to pay to their new landlord. It's easy to imagine the kinds of disputes that this would have caused.

There is some evidence that Tycho overstepped the mark in his treatment of the peasants. In 1597, when his enterprise on Hven was coming to an end because of the meddling of hostile courtiers close to the new king, Christian IV, strident accusations of ill-treatment were brought against him by the islanders. Indeed, the island's pastor served a gaol sentence on Tycho's behalf.

What is certain is that when Tycho's affairs on Hven were finally wound up, following his death in exile, in 1601, the peasants quickly took matters into their own hands. They wreaked their revenge on his bad behaviour in the shape of bricks and stones, which they spirited away to their own homes. Within a few short decades, what had once been the greatest observatory in the world had virtually disappeared.

During the last year of Tycho's life, when he was in the imperial city of Prague, he worked closely with the youthful Johannes Kepler. Recognising Kepler's brilliance, Tycho confided closely in him, but the pair also quarrelled ferociously over minor issues. Indeed, Kepler has been posthumously accused—probably baselessly—of trying to murder his great teacher. But today they are reconciled, and their effigies stand side by side at the busy intersection of Parléřova and Keplerova in downtown Prague. Needless to say, when the Stargazer II tour visited Prague in 2010, those imposing statues provided the perfect backdrop for a group photo.

. . . AND A PRODUCT OF A VERY DIFFERENT TIME

One of the characteristics of our astronomy tours, particularly in Europe, is that there's so much of interest that participants often find themselves switching between widely differing historical periods when travelling from one destination to the next. Thus, the tours take quite unexpected twists and turns. In a similar vein, I'd now like to take you on a detour through a very different era of European history, a detour that nevertheless has links with Tycho's time.

Fast-forward nearly 350 years from Tycho's death and we find another stargazer putting on a display of bad behaviour related to the times in which he lived. But if we can be amused by Tycho's youthful incompetence with a broadsword, it's impossible to be anything but serious about the context in which this astronomer conducted his work.

Ernst Zinner was a German astronomer who lived through the Nazi period. In 1933, when Hitler came to power, he was 47 years old and director of the Remeis Observatory, at Bamberg in southern Germany. He was also a professor of astronomy in Munich. As a high-profile academic living and working in the heartland of Nazi fanaticism, Zinner came under pressure to join the Party and—to give him some credit—he refused. However, his refusal seems to have had more to do with snobbery than a distaste for the Party's totalitarian ideals. Like many of his fellows in the privileged world of academia, he regarded the Party's membership as uncultivated and working class.

Scrutiny by Party officials followed, and, although he was described as 'unfriendly and arrogant', he was not identified as an enemy of the state. He seems, therefore, to have had an adequate working relationship with the Nazi regime, and it has been suggested by the US historian Robert S. Westman that this was because his ideals broadly coincided with theirs: the superiority of the German race, the need for a Greater Germany, the idea of war as the epitome of male virtue, the need for authoritarian government and so on. Most disturbingly, when the 1937 Four Year Plan for the universities referred to 'cleansing German higher education from liberalist and materialistic bonds', Zinner seems to have been in accord. The dark clouds of the Holocaust were already clear on the horizon.

The evidence for Zinner's sympathies comes from his own work. His speciality was the study of variable stars—whose brightness changes in a systematic way—but he was also a prolific writer on the history of astronomy, and in this it is possible to identify not just bad behaviour but an attempt to rewrite history to match Nazi ideals. In 1943—exactly 400 years after the astronomer Nicolaus Copernicus' epoch-making *On the Revolutions of the Heavenly Spheres* had appeared—Zinner published a book entitled *The Origins and Dissemination of the Copernican*

Doctrine. In this volume he gathered together a vast archive of biographical and other data on obscure individuals and their roles in promulgating the Copernican theory of the Solar System, which places the Sun rather than the Earth at the centre. No doubt it is the archive that gives the book its enduring value (it was republished in 1988), but what is more relevant to us is the accompanying text.

Zinner makes an outright attempt to turn the Copernican model into a German creation. Wherever possible he seeks a Germanic link with the events surrounding the development of Copernicus' Sun-centred model and its subsequent dispersion into the world of learning. Copernicus himself is turned into a German rather than a Pole by dint of dubious genealogical and geographical argument. More surprisingly, Tycho Brahe is transformed into a kind of Germanised Dane. It is true that Tycho spent a significant amount of time in Germany as a young man—indeed, his fateful duel was fought in Rostock—but it is hard to imagine him considering himself to be anything other than a Danish nobleman.

Tycho, in fact, was not a supporter of the Copernican theory, preferring his own hybrid model of the Solar System, in which the planets orbit the Sun, which in turn orbits the Earth. This does not seem to have worried Zinner, however. More of a problem for him was the company that Tycho kept. One of Tycho's colleagues late in his life was a Prague astronomer and mathematician called David Gans—a Jew. Gans held the view that natural sciences such as astronomy could have a unifying effect for Jews and Christians, bringing together their theologies in a beneficial manner. This was in line with Tycho's own thinking, and the two men spent much time together around the turn of the seventeenth century discussing the astronomy and astronomers of the past. References to this important figure in Tycho's life can be found in one of Zinner's earlier publications, but by the time his *Origins and Dissemination* was published—meticulously researched though it was—Gans had conveniently disappeared. Tycho's circle had been sanitised to meet the expectations of a readership looking for pure Aryan credentials.

No doubt one could argue that this kind of distortion was as much an effort at self-preservation as support for the Third Reich. It cannot have been easy for educated people to survive in Nazi Germany. And perhaps the fact that Zinner nowhere embarks on overtly anti-Semitic rhetoric in his book supports that view. However, there is other evidence from as early as 1931 of racist tendencies in Zinner's thinking. Having compared the improvements in stellar position measurements made by individuals whom he could identify as Germanic in origin with an earlier decline in their Greek equivalents, he suggests the foundation of an institute to investigate national tendencies in science and culture. It is a curious idea with unpleasant undertones.

After the Second World War, Ernst Zinner continued his work almost as if nothing had happened. He retired from his position at Bamberg in 1953 but continued his interest in rare astronomical literature until his death, in 1970. Most of his own 5000-strong collection of books was sold to the California State University at San Diego in 1967.

What are we to say about this stargazer behaving badly? That, compared with the events surrounding him, his behaviour was mild? Certainly, against the broad backdrop of Nazi oppression, Zinner hardly rates a raised eyebrow. And from our vantage point in the 21st century it is easy to be judgemental. But one would like to think that a true and dedicated astronomer might have done better. Much better.

WHO INVENTED THE TELESCOPE?

Now, back to the seventeenth century, and to an episode that is still surrounded by mystery and intrigue—and bad behaviour. The invention of the telescope was arguably the single most important event in the entire history of astronomy. For the first time, one of the human senses was extended to discover objects that had previously been invisible. It instantly transformed the study of the heavens from simple join-the-dots measurement to a dawning understanding of what those dots actually stood for. At a stroke it gave new weight to Copernicus' Sun-centred theory of the Solar System, beginning an era in which theories about our environment in space could be built on an ever-increasing scale. Who could have guessed that those ideas would one day lead to the supremely powerful Big Bang model of the Universe, revered by today's scientists and philosophers?

The telescope first came to prominence in the hands of Florentine patrician Galileo Galilei, who, at the time, was far from his native Florence, teaching mathematics in the University of Padua, near Venice. You have only to read a modern translation of his little book *The Starry Messenger*, of 1610, to sense the trembling excitement with which he plucked discovery after discovery from the sky with his new gadget. Craters and mountains on the Moon, stars rather than, well, milk coagulating in the Milky Way and—most dramatically—Jupiter's brilliant speck revealed as a disc with four moons orbiting around it. Truly, this was the dawn of modern astronomy.

Contrary to popular belief, however, Galileo wasn't the inventor of the telescope. He tells us in *The Starry Messenger* that he first heard a rumour about a Dutch perspicillum (spyglass) sometime around May 1609. The story of how Galileo set about understanding this device without actually having seen it and then went on to build a succession of steadily improving versions is the stuff of legend. Without question, the startling discoveries he made resulted directly from those efforts. But invent the telescope he did not.

So who did? It's a great question, and the short answer is that no one knows, exactly. What we do know is that truly extraordinary events surrounded the emergence of the telescope and they make for a remarkable story. Even today's historical research on the subject has so many twists and turns that it's worth a little scene-setting before we unpick the details.

The first name associated with the telescope's debut on the world stage is that of Hans Lipperhey (commonly spelled Lippershey), a humble spectacle-maker, who petitioned the government of the fledgling Dutch republic for a patent on the invention late in September 1608. His timing was perfect, coinciding with tense diplomatic negotiations between the Dutch and the Spanish, who had been at war since 1568. But, in a farcical turn of events, two other individuals appeared within three weeks asking for similar patents, so it is questionable whether Lipperhey was the true originator of the telescope.

Many of the contemporary documents that refer to these bizarre proceedings were uncovered early in the twentieth century by another Dutchman, Cornelis de

Waard, and presented, together with related evidence, in his *Invention of the Telescope*, published in The Hague in 1906. It was in assembling and translating those original sources for a wider readership that the modern historian Albert Van Helden performed perhaps the greater service to today's scholars. He wrote a detailed analysis of their contents in a comprehensive work also entitled 'The Invention of the Telescope', published as a complete issue of the *Transactions of the American Philosophical Society*, in 1977. This study has become the yardstick against which all subsequent commentaries on the origin of the telescope have been judged. It should not be assumed, though, that Van Helden solved all the problems. Indeed, he made no claim to have done so. The evidence is a maze of contradictory statements and reports, often with well-known historical names intermingled with shadowy figures in confusing circumstances. It's a very difficult research area.

Then, into this minefield stepped another brave author, the late M. Barlow Pepin, who, in 2004, produced a little book with the intriguing title of *The Emergence of the Telescope: Janssen, Lipperhey, and the Unknown Man*. His stated purpose was to take a 'fresh look' at the circumstances under which the telescope had arrived on the scene. While no significant new evidence had come to light since Van Helden's epic work, Pepin did manage to draw together some previously unrecognised threads, and his book is a useful contribution to the scholarly literature. To be honest, I wish I'd had a copy a few years before, when I embarked on my own foray into this morass for *Stargazer: The Life and Times of the Telescope*. The conclusion Pepin arrived at is not too different from de Waard's of a century earlier, although it is a lot more entertainingly presented. The telescope was perfected not by Lipperhey, said Pepin, but by one Sacharias Janssen, a spectacle-maker, peddler and small-time crook, who secretly presented an example to the authorities shortly before Lipperhey got around to it. It was a classic example of spectacle-makers behaving badly.

Several authors, including Van Helden, have asked another intriguing question: 'Did the telescope have a history before 1608?' Perhaps there was ancient knowledge of the working principles of the telescope that was lost in the mists of time, or maybe more-recent experiments by optically minded natural philosophers predated the appearance of the Dutchmen's telescope in the historical record. A few of the many claims to such precursors have some credibility, and one in particular stands out, in a book with the wonderful title *Natural Magic*, first published in 1589, and written by an Italian optician called Giovanbaptista Della Porta. A throwaway line about the use of lenses to correct poor eyesight quickens the pulse of anyone wishing to place the invention of the telescope before the early seventeenth century. Specifically, Della Porta refers to the use of convex lenses (which are thicker in the middle than around the edge) and concave lenses (which are the other way around):

> With a Concave you shall see small things afar off, very clearly; with a Convex, things neerer to be greater, but more obscurely: if you know how to fit them both together, you shall see both things afar off, and things neer hand, both greater and clearly.

Remarkably, when the telescope did finally make its undisputed first appearance, it was in a form that took a convex and a concave lens and 'fit them both together' at opposite ends of a tube. This makes what is called a Galilean telescope, in homage to its great champion, and, incidentally, is still found today in the guise of ordinary opera glasses.

So, did Della Porta know the secret of the Galilean telescope in the 1580s? Maybe he did, although it is clear from the context of the quote that he is really referring to the improvement of defective vision rather than a telescope in the modern sense of the word. Nevertheless, Van Helden has suggested that Della Porta did, indeed, succeed in making a weak Galilean telescope but that he didn't perceive its wider possibilities as a device for magnifying distant objects.

Credibility is lent to this idea by the fact that, on 28 August 1609 (by which time Lipperhey's Dutch spyglass of 1608 had become well known), Della Porta wrote to a noble friend, telling him that he'd seen one of these supposedly new instruments. 'It is a hoax,' he said, 'and is taken from the ninth book of my *On Refraction*.' By 'hoax', we can perhaps infer that Della Porta meant there was nothing new in the Dutch invention, for hadn't he written about it himself twenty years previously? In fact, Della Porta was getting his own books confused, since *On Refraction* (1593) doesn't mention the combination of lenses, whereas, as we have seen, *Natural Magic* does. I guess when you're a sixteenth-century genius you might have some excuse for mixing up your references. Notwithstanding that slip, the letter does seem to support a case for the Galilean telescope having originated in Italy, a theme to which we shall return.

The one thing that casts doubt on the effectiveness of this and other sixteenth-century supposed telescope-like devices is that they were not picked up for widespread use as military or scientific instruments, as they certainly were in 1608 (very quickly, in fact—the news spread like wildfire throughout the following year). In particular, the greatest scientist in Europe in the late sixteenth century—Tycho Brahe—seems to have been completely unaware of them. Had he heard of a telescope in any form, he, more than anyone, would have checked it out and would have either made one or written about it. But he did neither.

SEPTEMBER 1608

Six years and eleven months after Tycho's death, the telescope was a reality. That much is historical fact. We know, from a number of original documents that have survived the ravages of time (and a few that remain only as copies, thanks to the attentions of the Luftwaffe in 1940), that sometime around Saturday 27 September 1608 a spectacle-maker turned up in The Hague carrying a telescope and a letter of introduction written two days earlier.

As it happens, that was quite a weekend in Dutch history. With both sides weary of conflict after 40 years, the Dutch and the Spanish had taken tentative steps towards a truce, and peace negotiations had been in progress in The Hague throughout September. But things had not gone well, and on the last day of the month the commander-in-chief of the Spanish forces in the southern Netherlands—a nobleman by the name of Ambrogio Spinola—left The Hague with the talks in deadlock.

Spinola's opposite number on the Dutch side was the commander-in-chief of the armed forces of the United Provinces of the Netherlands, Prince Maurice of Nassau. (Not to be confused with its better-known namesake in the Bahamas, Nassau in present-day Germany was then a principality of the Holy Roman Empire.) Maurice was a seriously talented man, and pretty formidable in appearance, if contemporary portraits are anything to go by. Not only was he commander-in-chief, but he was also the Stadtholder (leader) of the United Provinces government, an assembly known as the States General. This federal parliament was made up of representatives of the seven provinces of the Dutch republic. On or around 27 September, the delegate in The Hague from the province of Zeeland received from home, two days' journey away, a strange letter from his fellow councillors. Dated 25 September, the letter explained that

> the bearer of this [letter] . . . claims to have a certain device, by means of which all things at a very great distance can be seen as if they were nearby, by looking through glasses which he claims to be a new invention, [and he] would like to communicate the same to His Excellency [Prince Maurice].

That the parliamentary recipient of the letter did, indeed, introduce the bearer of the letter to Prince Maurice is borne out by other contemporary documents. The Hague was full of diplomatic emissaries from other nations, all with a vested interest in the peace process, and at least one of them wrote home to tell of a humble spectacle-maker who had been ushered in to the prince's presence carrying a remarkable tube with a glass lens at each end. It was called 'the instrument for seeing far', for the word that became 'telescope' in English was not coined until three and a half years later, when an exclusive group of Italian and Greek intellectuals held a banquet to honour Galileo and his astronomical discoveries.

In order to test the instrument, Prince Maurice climbed the tower of his residence in the grounds of the Binnenhof—the imposing thirteenth-century building that was then the seat of the States General and remains its ceremonial home today. From there, we are told, he could clearly see through the instrument the clock of Delft and the windows of the church of Leiden, respectively one and a half and three and a half hours' journey away. The enormous military potential of the device was not lost on the Stadtholder, and he was *very* impressed: so much so that within a day or two—or perhaps even at the same time—he showed it to the enemy commander, Marquis Spinola. We don't know whether that was a misguided ploy or some sort of diplomatic etiquette—or maybe Prince Maurice just wanted to deliver a gloating 'look what *we've* got' to his opposite number. Whatever the reason, by the time the commander-in-chief of the enemy forces left The Hague, on 30 September, he had not only held the new surveillance device in his hands but taken a look through it. No doubt he was impressed, too.

On the following Thursday, 2 October, the spectacle-maker presented himself at the Binnenhof to be interviewed by the members of the States General, who were keen to see the marvellous new invention for themselves. He asked them to grant him a patent for 30 years, during which time no one else would be permitted to make telescopes. Or, if they preferred, he'd be quite happy with a yearly pension, in return for which he would make telescopes solely for the state. He didn't really mind which.

The minutes of that interview give us the identity of the man for the first time, since the councillors' letter of 25 September didn't actually mention his name. He was, as we have seen, Hans Lipperhey, a German-born spectacle-maker who lived in the town of Middelburg, in

Zeeland—and he was the first person to be identified with a verifiable working telescope. He didn't get his patent, but he was paid handsomely to make three binocular versions of the telescope, the last of which he delivered on 13 February 1609. Whether or not he invented the telescope itself, Lipperhey can safely be hailed as the inventor of binoculars. And that was a task so difficult that over the next two centuries only a handful of instrument-builders even attempted it.

Hans Lipperhey's moment of glory as the inventor of the telescope lasted little more than two weeks. News of his wonderful device spread quickly through the provinces of the Netherlands and evidently caused consternation in at least two Dutch households when it reached them.

On Friday, 17 October, a letter arrived at The Hague addressed to the States General from an instrument-maker in Alkmaar, in the northern part of the Netherlands, a man named Jacob Adriaenszoon but more commonly known as Jacob Metius. He was a much more imposing character than the humble Lipperhey. His father, Adriaen, was a former burgomaster of Alkmaar, and his brother, another Adriaen, was a professor of mathematics and astronomy, who had studied with none other than Tycho Brahe. (Such was the extent of Tycho's influence throughout Europe at the time.) The letter was a long, rambling missive announcing Metius' accidental discovery of the principle of the telescope after a couple of years of intensive study and experimentation. It went on to say that Metius had heard of the invention of the spectacle-maker of Middelburg and that his own prototype instrument had been tested and was at least as good as that one, so he, too, deserved a patent on the invention, because of his own 'ingenuity, great labour and care (through God's blessings)'.

Metius' petition was, like Lipperhey's a fortnight earlier, duly noted in the States General's minute book, with a comment that Metius would be granted modest funding and 'admonished to work further in order to bring his invention to greater perfection, at which time a decision will be made on his patent in the proper manner'. With hindsight, this seems rather generous of the States General.

Hardly was the ink dry in their minute book than yet another claim turned up. It may even have arrived before Metius' petition, although was not minuted then. Written on 14 October, this was a letter from the same councillors of Zeeland who had provided Lipperhey's original letter of recommendation. In Middelburg, they said, there were now others who knew the art of seeing far things and places as if nearby. In particular, there was a young man who had demonstrated a similar instrument to Lipperhey's. What would the honourable gentlemen like them to do about him? If there was any doubt left in the minds of the parliamentarians in The Hague that this invention was already too widely known to be patented it must have been dispelled altogether by this news.

With those three players assembled on the stage of history, the scene was set for the true details of the telescope's origin to disappear amid the babble of conflicting contemporary records. Who first made a telescope? Lipperhey? Metius? Or the 'young man' of Middelburg? And who was that last young rascal, anyway?

In fact, 'rascal' is probably an appropriate epithet. Since 1656, when his association with the first telescope was confirmed by the recollections of several locals who had known him, most scholars have taken the young man to be Sacharias Janssen, a spectacle-maker who would have been about twenty years old in 1608. But it was Cornelis de Waard, in the early twentieth century, who uncovered most of what we know about him. And it makes pretty interesting reading.

As well as following the family trade of spectacle-making, Janssen also peddled his wares throughout Europe. He may have been the individual who was reported as attempting to sell a telescope with a cracked lens at the Autumn Fair in Frankfurt in September 1608. He was, in any case, frequently on the wrong side of the law, with assault and non-payment of debt among the many items on his criminal record. Most serious were two accusations of forgery, for which he was eventually threatened with the death penalty. It seems that he had produced counterfeit Spanish coins in an admirably patriotic attempt to undermine the Spanish economy but somehow had forgotten to stop when the Spanish and Dutch eventually signed a truce in 1609. Wisely, Sacharias disappeared before the sentence could be carried out.

Other records show that the premises of the Janssen family business, in the market square near the abbey in Middelburg, were just a few doors away from Hans Lipperhey's. Perhaps ideas had flowed from one to the other. Even more interesting is the testimony of a man who claimed to be Sacharias Janssen's son—one Johannes Sachariassen. In 1634, Sachariassen boasted to a friend that his father had constructed the first telescope in the Netherlands. He had copied it as early as 1604 from one that had belonged to an Italian, an instrument that had apparently carried the inscription '*anno 1[5]90*'. Albert Van Helden has presented a plausible scenario in which this mysterious Italian instrument may have been one of the weak telescopic aids to vision that Giovanbaptista Della Porta described in his *Natural Magic*, in 1589. Middelburg is known to have hosted large numbers of Italian exiles, most of whom were deserters—mercenary soldiers tired of helping the Spanish in their attempt to subdue the United Provinces. Perhaps Janssen did, indeed, make a copy of an instrument brought from Italy and managed to improve it to the stage that it could provide useful magnification.

In any event, these accounts suggest that Lipperhey might not, after all, have been the original inventor of the telescope, despite having been first off the mark in the historical record. Perhaps the proximity of his shop to Janssen's allowed him to gain some understanding of what his fellow optician was up to. There is even a suggestion in some sources that Lipperhey's children may have learned the secret of the telescope by holding up two lenses of the right kind, one behind the other—either by accident or after being shown.

CONSPIRACY THEORY

As we have seen, there is an enormous amount of roughly contemporary documentation on the origin of the telescope, dating from before 1608 to beyond the middle of the century. Much of it is confused and contradictory, and the problem lies in knowing how much weight to place on each item. In a chapter such as this, it is only possible to scratch the surface. However, the recent study by M. Barlow Pepin explores this material in great detail and arrives at a rather more subtle interpretation of the evidence.

Yes, says Pepin, Janssen did indeed perfect the telescope on the basis of a device he was shown by an Italian in 1604. But Lipperhey's knowledge of how to build such an instrument did not come merely from hanging around outside Janssen's shop doorway when things were quiet. It came, instead, from a much more intriguing source. Pepin cites an extraordinary passage in a book with the catchy title of *Telescopium, sive ars perficiendi novum illud Galilaei visorium instrumentum ad sydera*, by one Girolamo Sirtori, published in 1618 but probably written six years earlier when the events were still fresh in the author's memory. The book's title loosely translates as 'About Telescopes: Perfecting the Art of Making Galileo's New Instrument for Looking at the Stars', and in it we read that

> a man as yet unknown, with the appearance of a Hollander, visited Johann Lippersein [*sic*] at Middelburg in Zeeland. This [Lipperhey] is a man of striking air and appearance, and a maker of spectacles; as no-one else in that city: he [the visitor] ordered several lenses made, convex as well as concave. On the appointed day, the stranger returned, demanding the finished work. When he had them to hand, he picked up a pair; specifically a convex and a concave, placing one and then the other before his eye, separating them from one another, little by little as if to reach the focal point; after the manner of checking the spectacle-maker's handiwork. He then paid the artisan and left. The optician, who was not without inkling through all of this, began to do the same thing out of curiosity. He soon solved the problem which nature presented by fixing the spectacle lenses in a tube. As soon as he was finished, he hurried to the Palace of Prince Maurice and presented his invention. The prince already had a telescope, suspected that

> it might have military value, and of necessity had kept it a secret. But now that he found the concept had become known through chance, he dissembled, rewarding the industry and benevolence of the maker.

Pepin (from whose book the above translation is taken) goes to great lengths to identify the 'unknown man' mentioned in the passage. Remarkably, he finds evidence suggesting that it may have been an undercover emissary of the Spanish commander-in-chief Ambrogio Spinola, perhaps looking for the raw material to duplicate an amazing far-seeing instrument that his enemy, Prince Maurice, already had.

This passage brings into question the idea that Prince Maurice hadn't seen a telescope until Lipperhey delivered one to him. We are explicitly told that he had one already (or, in an alternative translation of the same passage, by Van Helden, *knew* of one already). So were there, in fact, two telescopes? Here, Pepin cites other evidence, specifically the work of a French diarist of the era, Pierre de L'Estoile, a minor government official with a keen eye for detail in contemporary events. In a diary entry for 30 April 1609 he describes telescopes he had seen on sale in Paris and adds

> I have been told that the invention was due to a spectacle-maker of Middelburg in Zeeland, and that last year he presented to Prince Maurice two of them with which things that were three or four miles distant were seen clearly.

This second mention of two telescopes lends weight to the idea. Of course, it's possible that de L'Estoile was simply misinformed, but taking the diary entry at face value it certainly supports the strange passage in Sirtori's book.

Pepin develops a scenario that accounts rather neatly for this evidence. Suppose the shady Sacharias Janssen had, indeed, spent four years developing the weakly magnifying device shown to him by an Italian and had managed to make a telescope. Suppose he had presented his invention to Prince Maurice rather earlier in the peace negotiations with Spain, no doubt with an eye on a patent. Now imagine that it was *this* instrument, not Lipperhey's, that Maurice gloatingly showed to Spinola. This is consistent with the historical documentation, but, significantly, it eliminates the need for all the action in the story to be compressed into the last two or three days of September, as it would have to have been if Prince Maurice had seen only Lipperhey's telescope. This is an aspect of Pepin's theory that is particularly appealing.

Pursuing the idea further, what does Spinola now do? He knows the telescope came from Middelburg. Wishing to obtain one for his own forces before he leaves the Netherlands, he instructs his undercover agent to go to the town, hunt out the person who made it and either buy one or buy the lenses to make one. But there's an unexpected problem. The itinerant Janssen is on his way to the important Autumn Fair in Frankfurt, 600 kilometres to the south, where he subsequently attempts to sell a telescope with a cracked lens. So when the undercover agent looks for a spectacle-maker in Middelburg, whose shop does he find? The passage in Sirtori's *Telescopium* gives us the answer: not Janssen's, but Lipperhey's, a few doors away.

Lipperhey, having been given the secret of the telescope by Spinola's undercover agent, quickly knocks up one of his own, thinks, 'I might be able to earn a fast buck or two from the government with this', and heads off towards The Hague, having first obtained a letter of introduction from one of his local councillors. And, because Prince Maurice doesn't say 'Oh, I've got one of those already' when Lipperhey eventually arrives to present his telescope to him, the official record accords priority in the invention to Lipperhey.

Pepin suggests that the reason for Prince Maurice's secrecy is at least partly due to Janssen's shady character, the spectacle-maker much preferring clandestine dealings to anything open and above board. And, indeed, that might also have suited the delicacy of the diplomatic situation. But things came unstuck for Janssen when he arrived back in Middelburg after the two-week trip home from Frankfurt and found that Lipperhey had somehow pre-empted his claim on the invention. Perhaps Janssen's wish to regain his priority without giving too much away explains why the second letter from the Zeeland councillors to the States General speaks only of 'a young man', without mentioning any names.

Whatever your view of this interpretation of the evidence—unbelievably far-fetched or entirely plausible—I hope you will agree with me that Pepin has performed a service to historians of the telescope by suggesting possibilities that are far from obvious. That, in itself, is a step forwards.

For what it's worth, however, my own opinion is that this scenario still leaves questions unanswered. I find it hard to believe that Prince Maurice would not have given some hint of the turn of events to the members of the States General, who, unless they were particularly disingenuous in their record-keeping, were completely unaware of the existence of the telescope before Lipperhey appeared. Moreover, the documentary evidence is hardly watertight, being based entirely on second-hand reports.

I suspect that we will never know for certain who actually invented the telescope, unless completely new evidence appears. The answer to the question of why it emerged so suddenly from obscurity, apparently in several places at once, might simply be that the necessary ingredients were all there. The skills of glass-makers and opticians had recently developed to the stage where lenses of the required quality could be produced. Once they were available it was inevitable that someone would eventually stumble across the right combination of lenses. And, with the Netherlands embroiled in military crisis, you'd have to be pretty dumb not to perceive the possible value of such an invention. Competition from other spectacle-makers prone to bad behaviour would only pile on the pressure.

Clearly, several people knew the secret of the telescope in the early 1600s. Metius' claim to originality appears to be genuine, but Alkmaar was relatively remote. Middelburg, on the other hand, had a major glass factory of its own, the only one in Zeeland. That small Dutch town seems the most likely place for the first real telescope to have emerged from the work-stained hands of its maker.

As we have seen, Galileo transformed the telescope into an engine of discovery for astronomy. But it would be wrong to imagine that he was the first to point a telescope towards the sky. In a note written in October 1608, one of the foreign diplomats observing the peace process said of Lipperhey's telescope that 'even the stars, which ordinarily are invisible to our sight and our eyes, because of their smallness and the weakness of our sight, can be seen by means of this instrument.' Someone, therefore—we have no idea who—had noted the telescope's ability to enhance the eye's sensitivity to faint light early in the piece, perhaps even during those last few days before Spinola's departure from The Hague at the end of September. Notwithstanding all the questions that still surround the telescope's appearance—and all the bad behaviour that went with it—there can be no doubt that that extraordinary month in the northern autumn of 1608 truly heralded astronomy's awakening.

SEPTEMBER 2008

If you're going to have a birthday party, the one thing you definitely need is a cake. But our cake was beginning to look like only a remote possibility. You couldn't say it was anyone's fault, though. A few days before, there had been a fire in the Channel Tunnel, and all Eurostar trains out of London were still hopelessly behind schedule. Having risen at the crack of dawn we'd had an interminably long wait at St Pancras, and it was only because our Stargazer tour coordinator, Marnie, and her offsider, Emma, had managed to find the right person in the Eurostar office to flutter their eyelashes at that we had got onto the train at all.

So our arrival in Brussels was much, much later than planned. The birthday party was scheduled to take place that evening in The Hague, 400 years to the day (almost) since Hans Lipperhey had arrived there with his telescope. It was to be a grand affair, as befits a party for something so important. But the cake-makers to the Dutch royal family, from whom, months earlier, Marnie had ordered the birthday cake, closed their doors strictly at 6 pm. And The Hague is a two-hour coach ride from Brussels.

In the event, we made it. But our arrival in The Hague, at 5.55 pm, was more akin to a space shuttle re-entry than the dignified entrance of a study tour into a provincial capital. At a spot that gave every appearance of being a freeway intersection in the middle of nowhere the coach screeched to a halt, and Marnie shot out and disappeared into the distance, taking hurried directions from the folk at the cake shop via her mobile phone as she ran. Piecing things together afterwards, I guessed we were where the E19 joins the A4, but who knows? The bottom line was that I never expected to see Marnie again.

Astonishingly, she got there in time to pick up the most beautiful cake specially decorated with a seventeenth-century telescope made of chocolate. All was well. All she had to do was pay for the cake and catch a cab to our hotel. Ah, but no. 'Unfortunately, we don't take credit cards . . .' With Marnie's wallet full of pounds rather than euros and no sign of an ATM, this looked like the end of the road. But another five minutes of fast talking found Marnie—a stranger from the other side of the world—walking out of the door carrying a €210 cake and leaving behind only the vaguest promise of how it would be paid for.

The party was an outstanding success and a highlight of the entire tour. We toasted the telescope and all its fans—Hans, Jacob, Sacharias, Maurice, Ambrogio and even the 'unknown man'—late into the night. The one person who missed out was Galileo, but that was only because the following week we were heading for his old stamping ground in Padua, where he'd perfected the telescope, and later for his home turf of Florence. So Galileo would be well toasted, too.

The morning after the birthday party, despite being slightly the worse for wear, a small group of us were heading down to Middelburg, where a symposium on the invention of the telescope was being held. Some good friends of mine from the world of telescope history were going to be there, as well as scholarly megastars like

Albert Van Helden, so I was keen for us to be on our way. Somehow, though, the same gremlin that had plagued the trip from London had been busy again.

Marnie, having picked up a rented minibus for the trip from a local garage, accepted their help in programming the GPS to get her back to her waiting passengers at the hotel. But the incorrect destination that they accidentally entered sent her off on the motorway to Rotterdam—from which there's no easy exit. An hour and a half later she arrived to pick us up, just a tad flustered. 'Oh, don't worry,' we said. 'We can set up the GPS to get us there in double-quick time.' One of the group punched in the details of the Roosevelt Academy, where the symposium was being held. Two hours later, while we circled aimlessly through an anonymous housing estate on the outskirts of Middelburg, the GPS confidently informed us that we had arrived.

The symposium, when we finally did arrive, was truly inspiring. It was great to chat to some of the experts in the field, and I was quietly delighted when a few of them said they had read some of my own efforts on the subject. And then it was almost time to head back to The Hague—but there was one more pilgrimage to make first. The abbey church in Middelburg, whose imposing façade had separated the shops of spectacle-makers Hans Lipperhey and Sacharias Janssen at the turn of the seventeenth century, is still there. Rebuilt following extensive damage during the Second World War, the area reveals no trace of the shops that once dotted the square, but the church has been beautifully restored and stands in imposing defiance of the ravages of war. For anyone inspired by the story of the birth of the telescope, this is hallowed ground, and we relished our few moments there before continuing on our way.

Mercifully, the trip back to The Hague was uneventful, but once Marnie had dropped off all the tired and grateful telescope nuts at their hotel there was one final task for her. An early flight to Copenhagen the following morning meant that this had to be done *now*. So, stealing out into the night, she drove the empty minibus to somewhere near the junction of the E19 and the A4, found the shop of the cake-makers to the Dutch royal family and stuffed €210 under the locked door. I wonder if they ever found it?

A PRODUCT OF HER TIME?

There's one last piece of bad behaviour I'd like to relate, and it concerns our tour destination immediately before the journey to The Hague. We were on another pilgrimage, this time to Greenwich, where the Royal Observatory was founded by decree of King Charles II, in 1675, 'for the perfecting of the art of navigation'. Walking around Christopher Wren's beautiful Octagon Room, purpose-built for astronomical observations, in the leafy surroundings of Greenwich Park, it was easy to imagine the first astronomer royal, John Flamsteed, carrying out his work there.

The bad behaviour I am referring to was not on the part of Flamsteed—who, like Tycho, did have a short fuse—but on the part of his wife, Margaret. Not wishing to appear in any way discriminatory, you see, I looked for examples of female astronomers behaving badly when I was preparing this chapter. I'm afraid I looked in vain. Partly, of course, that is because until recently there have been few women astronomers, but I suspect it also says something about the equanimity of women under duress.

The best example I could find was that of Margaret Flamsteed. There's no doubt that she had an exemplary record as a devoted wife and the perfect hostess when visitors came to the observatory. But when she became a widow, in 1719, she found herself faced with the monumental task of completing two of the three volumes of the *British Catalogue of the Heavens*, Flamsteed's great catalogue of star positions, which had been in preparation when he died. She engaged the help of her late husband's assistant Joseph Crosthwait and a former colleague, Abraham Sharp. You may remember Sharp as the owner of Horton Hall in Yorkshire, whose crumbling shell I visited as a youngster. Horton Hall was 300 kilometres from Greenwich—half a world away, in the 1720s. But despite this great inconvenience, Sharp and Crosthwait managed between them to complete the work.

The three volumes eventually appeared in 1725, dedicated by Margaret Flamsteed to King George I, but without any mention of the laborious efforts of Crosthwait and Sharp. Worse followed. When Mrs Flamsteed died, four years later, Crosthwait wrote bitterly to Sharp at his Yorkshire home:

> You and I have laboured for Mrs Flamsteed for above ten years, and our reward, so often promised, is at last befallen us. Mrs Flamsteed died the 29th ultimo, and has given . . . to you and I not one farthing . . . Could Mr Flamsteed have seen her ingratitude, I am confident he would not have left it in her power.

It's a sad letter, but them's the breaks. It's all you can expect when you deal with stargazers (or their spouses) behaving badly.

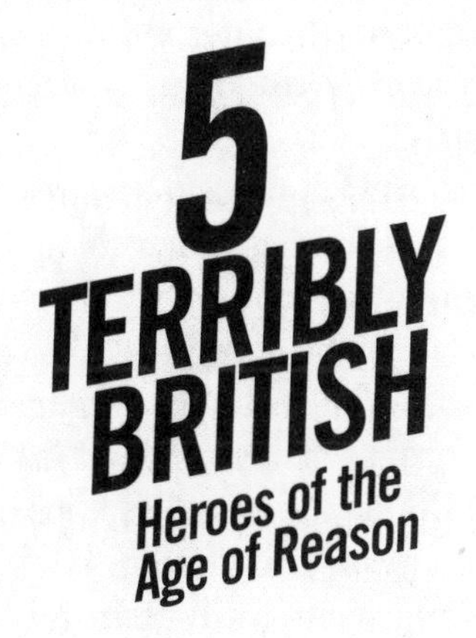

It's hard to avoid the Universe these days. It's everywhere—if you'll pardon the pun. Newspapers and TV news bulletins feature the latest images from the Hubble Space Telescope and the exploits of robotic deep-space probes. Astronomy magazines vie for your attention in the local newsagent's (usually among the New Age and astrology exotica—but at least they're there). Radio stations run stargazing segments from both professional and amateur astronomers. And stunning TV documentaries wow you with the wonders of the Universe. It's all terrific—and long may it continue.

It was not ever thus, however. Back in the 1950s, in drab, post-war Britain, you had to look hard in a library to find a book on astronomy, let alone see any media coverage. There was none of the exciting in-your-face science that we are so used to today. And absolutely nothing was in colour—with the single exception of the *Eagle* comic and its science-fiction hero, Dan Dare (whose exploits in space contained a good measure of real science, thanks to the expertise of his gifted creator, the late Frank Hampson).

So you can probably appreciate the inspirational effect on young would-be scientists that a colourful and authoritative production on astronomy would have had when it appeared. But you may be hard pressed to imagine where you would have found it. A book? A magazine? Neither. It was in—wait for it—packets of tea. *Tea?* To be precise, it was in Brooke Bond Choicest, PG Tips or Edglets, leaf teas all, with not a teabag to be found. It's hard to imagine anything more eccentrically British than astronomy with a cup of tea thrown in. Or, rather, poured in.

This curious venture came about when the Brooke Bond Tea company, convinced of the sales potential of collectable cards in its products, commissioned a series of 50 informative picture cards on astronomy called 'Out into Space'. A quarter of a century earlier, cigarette companies had pioneered the card-collecting craze, but astronomy tea cards were a new thing. The ploy worked, and the cards were avidly collected by the nation's ten-year-olds. Quite a few of today's best-known astronomers had their careers decided by Brooke Bond's seductive artwork. One such was David Allen, a truly gifted astronomer and science communicator at the former Anglo-Australian Observatory in Sydney, who sadly died in 1994. Towards the end of his life, David spoke with affection of the tea cards that had inspired him as a boy in Manchester. And you may not be surprised to hear that there's a bookcase close to my keyboard that contains a complete set of the cards, secure in the album that could be bought 'from your grocer, price sixpence' to house them. Yes, they inspired me, too, and are today a treasured keepsake.

Each card in the series has colour artwork on one side and an explanatory text on the other, printed, of necessity, in a microscopic type. The identity of their author is not recorded, but they do carry the blessing of a prominent astronomer, the late Alan Hunter, whose affiliation is noted on the cards only as secretary of the Royal Astronomical Society, although he became acting director of the Royal Greenwich Observatory in the mid-1970s. By then, that venerable institution was no longer in Greenwich, having long before relocated to the darker skies of Sussex. I remember Hunter masterminding its 300th birthday celebrations, in 1975, with flair and skill, when I was a youthful staff member there. In his role as boss of the observatory, Hunter was following in the hallowed footsteps of John Flamsteed and ten subsequent astronomers royal. In 1972, however, the positions of astronomer royal and director of the observatory had been separated. It was the beginning of the end—a little more than a quarter of a century later, the Royal Greenwich Observatory ceased to exist.

A browse through the tea cards with today's hindsight is an interesting exercise. You get the impression of an arbitrary mishmash of rather old-fashioned astronomy mainly highlighting the planets and stars. Galaxies—the immense objects that we now know contain hundreds of billions of stars—are called 'spiral nebulae', a term that was already out of date by 1930. More recent headline-grabbers, like quasars, black holes and neutron stars, were unknown, of course. Apart from one picture of a radio telescope, the cards could have been produced a century ago, and the fact that Pluto barely rates a mention has less to do with our modern understanding of its place in the Solar System than the cards' author not having got used to its presence in the first place.

Several cards cover basic astronomical phenomena like the seasons, eclipses and tides, although the poor old Moon receives only a background sketch in the album. The planets each have a card of their own, and there are no fewer than three describing the signs of the zodiac. Another three depict a ragbag collection of astronomical instruments: the radio telescope, a glass prism for analysing starlight and a sixteenth-century astrolabe—in that order.

The bulk of the set, however, is devoted to pictures and descriptions of the constellations. Twenty-two cards cover an odd mixture that leaves you wondering quite what the selection criteria were. Orion, for example—the brightest and most recognisable constellation in the entire sky—didn't make it. In spite of that, it is in the constellations that the real charm of the cards emerges. Colourful star maps with outline figures of their mythical counterparts set against a midnight-blue sky evoke something of the magical appeal of simple, naked-eye stargazing. They give the cards the air of belonging to the Age of Enlightenment, when science was just beginning to disentangle itself from the beliefs and superstitions of former times.

Quaint though they seem today, the tea cards were truly the best thing around for a 1950s Pommie youngster interested in astronomy. All the coolest kids in school had them, and parents went frantic buying up packets of tea, trying to find the last few missing cards. But the cards also had an interesting side-effect—for me, at least. To this day, I can't smell the contents of a packet of tea without getting a strange feeling that there's something exciting hidden at the bottom of it. I wonder if anyone else has experienced that sort of thing? While you're thinking about it, I'll just go and put the kettle on.

IN THE FACE OF THE SUN

The United Kingdom, of course, is a Mecca for astronomy history nerds. In many respects it is where telescopic astronomy came of age, following its birth in the hands of Galileo and his immediate successors. Early discoveries by Continental astronomers during the second half of the seventeenth century, such as cloud belts on the giant planet Jupiter and the misty patch of light we now know as the Andromeda Galaxy (discovered by Neapolitan and German astronomers respectively), led to significant advances in our understanding of the heavens. Prominent among these astronomers were Johannes Hevelius, Christiaan Huygens and Giovanni Domenico Cassini. A Polish brewer, a Dutch nobleman and an Italian working in Paris—it sounds like the start of a bad joke. But British telescopic astronomy was quick off the mark, with a fellow called Thomas Harriot sketching the lunar surface from London as early as July 1609, months before Galileo had perfected his own telescope.

Even more spectacular—albeit pitifully brief—was the career of a brilliant young Lancastrian named Jeremiah Horrocks. The son of a Liverpool watchmaker, Horrocks was born in 1618 and entered Cambridge University fourteen years later. There, his youthful interest in astronomy flourished, and he developed a particular enthusiasm for the work of the great German mathematician Johannes Kepler, then only recently deceased. Although a great admirer of Kepler, Horrocks realised there were gaps in the great man's understanding of the mechanisms of the Solar System. In particular, the younger man made significant strides towards an understanding of the role of gravity a decade before gravity's greatest exponent, Isaac Newton, entered the world.

To Horrocks goes the credit for being the first person to realise that the Moon, like the planets, has an orbit that is not circular but slightly elongated into an elliptical shape. His careful observations and accurate computations even allowed him to estimate the amount of that elongation—a tricky operation from our vantage point at the centre of the Moon's orbit. He also recognised that the Sun influences the direction in which this elongation lies. But Horrocks' greatest triumph came in 1639, when he was only 21. By then, he was living in the Lancashire village of Hoole, working not as the curate of the local church, as is often stated, but in some unknown occupation. Horrocks' modern-day biographer, Peter Aughton, has suggested that his most likely position was tutor to the children of a wealthy local family. Whatever his job, Horrocks had time to pursue his astronomical work, and he became interested in an issue that had intrigued Kepler—the passage of Venus across the disc of the Sun. We now know that these events—transits of Venus—occur in pairs separated by eight years, with the intervening gaps alternating between 105.5 and 121.5 years, making a 243-year repeating pattern. The transits always occur in June or December (hence the half-years), and the early 21st century saw millions flocking to the world's observatories to witness the June transits of 2004 and 2012. With smartphone apps available to record it, the 2012 transit became a mass media event. If you missed it, your next opportunity won't be until December 2117. By then, you may well have lost interest.

Horrocks knew that Kepler had predicted the occurrence of a transit of Venus in 1631 but had died the year before the event, and that he had also said there wouldn't be one in 1639. But Horrocks guessed that the transits always occurred in pairs, and he began to make careful observations of Venus as it moved slowly through the sky towards its conjunction with the Sun during 1639. From his measurements, he calculated that the planet would, indeed, pass between the Earth and the Sun on 24 November—as reckoned in the Julian, or Old Style, calendar, still in use in Britain at the time. (Despite having long been superseded throughout much of Europe by today's Gregorian calendar, the Julian calendar was current in Britain until 1752. Well, what else would you expect?) The date of the transit in the Gregorian, or New Style, calendar was 4 December. Either way, it was a Sunday.

Forewarning a friend named William Crabtree of the impending transit, Horrocks made preparations to observe it by projecting the Sun's image onto a screen with his telescope, thus avoiding the need to look directly at the Sun through it—which is always dangerous. Crabtree lived some distance away, near Manchester, but had his own telescope, so he, too, set it up for projection. The transit was predicted to begin late in the afternoon, with the Sun low on the horizon on that short winter's day. Just before the due time, Horrocks was called away for some reason—possibly a church service—but when he returned, the small black disc of Venus was clearly visible on the Sun's image. He was able to record its slow passage across the solar disc for only a short time before the Sun set, but he was absolutely delighted. That delight was well founded—he had become the first person ever to witness such an event by using the newly invented telescope, and he had correctly predicted its occurrence himself. Crabtree, too, was successful, but only by dint of a break in the clouds just before sunset.

The significance of this for astronomy was in demonstrating the effectiveness of careful observation combined with refined calculation. All done by hand, of course—there were no calculators of any kind in those days. It improved our understanding of the orbit of Venus and, along with other measurements, led Horrocks to conclude (correctly) that the Solar System was much, much bigger than anyone had guessed. It was another half-century, however, before Edmond Halley (of comet fame) presented a scholarly paper to the Royal Society of London on the idea that Venus transits could be used to gauge the Earth's distance from the Sun. This eventually created the scientific imperative that drove Captain James Cook to Tahiti in 1769—and on to New South Wales.

What became of Jeremiah Horrocks? A few months after the transit, he returned to his family home, at Toxteth—today a troubled inner-city suburb of Liverpool, but then a nearby village. There, he further developed his astronomical theories, corresponding regularly with Crabtree and a handful of other astronomers in the north of England. Notable among them was a Yorkshireman, William Gascoigne, who had invented a clever device that allowed telescopes to be used for accurately measuring the separation of close pairs of objects such as double stars.

One might imagine that this cluster of bright young scientists could have become the nucleus of a learned society—a forerunner, perhaps, of the Royal Society, which came into being two decades later, in 1660. But it was not to be. On 3 January 1641, little more than a year after the transit of Venus, Horrocks died, at the age of 22. We don't know the cause of his death; Aughton suggests a heart problem but admits this is only a guess based on the lack of any other evidence. Robbed of the shining star of their little group, Crabtree and the others were heartbroken. They kept in touch with each other and endeavoured to gather together Horrocks' papers and letters.

But even worse events were about to befall them. Political and religious unrest had been smouldering throughout Britain during the 1630s, and, finally, in 1642, the nation exploded into civil war. The deep divisions that had long been growing cut through national and local boundaries, and through the educational and religious establishments, and even separated friends and families. Horrific battles followed, between the Royalist supporters of the monarchy, in the person of King Charles I, and the Parliamentarians, under Thomas Fairfax and Oliver Cromwell. They were the most brutal ever seen on British soil, inflicting terrible casualties, along with widespread looting and destruction. The Battle of Marston Moor, near York, on 2 July 1644, was especially punishing for the fledgling northern-English scientific community, most of whose members had Royalist sympathies. In short, the little group was decimated. Crabtree died after the battle, probably as a result of injuries sustained in it. Other Royalist friends fell, and, while Gascoigne survived Marston Moor, he died early in 1655, at Melton Mowbray. He was 24.

The civil war culminated, in 1649, with the defeat of the Royalists and the establishment of a republic, whose first act was the public execution of Charles I. Thus, Britain entered the puritanical era of the Commonwealth, in which scientific endeavour, like most other intellectual activities, was suppressed. With Horrocks and so many of his contemporaries gone, there was every chance that his work would be lost completely to posterity.

That, however, was not what transpired. Slowly at first, but with gathering momentum, the climate changed. With the death of Oliver Cromwell, in 1658, and the nation weary of puritanical impositions, the first steps were taken towards constitutional modifications that would allow the monarchy to be restored. They culminated, on 23 April 1661, in the coronation of Charles II. By then, men of science had again begun to gather. Indeed, on the very day of the coronation, a small group of astronomers, including the great Christiaan Huygens, assembled in London to observe a transit of the planet Mercury across the Sun's disc—a more frequent and less significant event than a Venus transit. In the early years of the Enlightenment, Charles II became a staunch supporter of cultural advancement and bestowed his patronage on the infant Royal Society in 1662.

It was that learned body which eventually restored Jeremiah Horrocks to his rightful place in British astronomy, although, curiously, the society was not the first to recognise his achievements. In the same year that it had received its royal charter, the Polish astronomer Johannes Hevelius published a manuscript written by Horrocks with the title *Venus in the Face of the Sun*. It was a detailed description of Horrocks' observations of the Venus transit and the calculations involved—characteristically enlivened with some of his own poetry. How the manuscript reached Hevelius a generation after Horrocks' death is an intriguing story in itself, especially since all the circumstances had seemed destined to condemn the work to oblivion. In fact, an undergraduate friend of Horrocks', John Worthington, initiated the rescue of his papers through his own recollections of the young astronomer's pre-war activities. With the evident brilliance of this unknown British scientist brought decisively to their attention, the members of the Royal Society resolved to publish all his surviving work. Despite the turmoil resulting from London's double whammy of the plague epidemic of 1665 and great fire of 1666, the society issued Horrocks' complete *Posthumous Works* in 1672, with the enthusiastic support of a young John Flamsteed, among others.

At last, the writings of the man who has since been called the 'father of British astronomy' became accessible to a wide readership. They were greeted with enthusiasm throughout the scientific world. And we can only wonder what the course of science might have been had this astonishing figure survived beyond the tender age of 22. It's no exaggeration to say that the greatest name in British science today could easily have been Jeremiah Horrocks instead of Isaac Newton.

THE AGE OF UNREASON

Not surprisingly, you would look in vain for Horrocks among the tea cards. More unexpectedly, you'd also look in vain for Newton. But at least the great man does figure among the background sketches decorating the album. He figures twice, in fact. On the first page there is a poor likeness of the youthful Newton, with a caption mentioning his discovery of the law of gravitation. Further along there's a rather more creditable drawing of the stumpy little telescope he made in 1668, usually taken to be the first to use a mirror rather than a lens to form the image. In that regard, Newton is often hailed as the 'father of the reflecting telescope'. But that paternity could be in doubt.

In Newton's time, telescopes for astronomy were like the old-fashioned draw-tube telescopes we're all familiar with today: a long tube with a glass lens at each end, and maybe one or two inside as well. In the quest for a better view, however, these telescopes had been taken to extremes, becoming steadily longer throughout the seventeenth century. The reason for this lay in a fundamental flaw in all early refracting telescopes. They suffered from chromatic aberration, a colourful defect resulting from the front lens splitting light into unwanted rainbow hues. This problem was eventually solved by combining lenses of different glasses, but in Newton's day the only known remedy was to make telescopes long and thin—and the longer they got, the more unwieldy they became.

Ridiculously long telescopes dominated the astronomy of the time, and Christiaan Huygens was one of the chief offenders. Another was Johannes Hevelius, who built a series of instruments that culminated, in about 1670, in one whose length was no less than 46 metres. In its construction, this staggering contraption had more in common with the rigging of a sailing ship than an optical instrument. From a mast 27 metres high, the telescope's 'tube' of planks was suspended by ropes and pulleys. It is easy to imagine the chaos that would have faced anyone trying to observe with it on a dark night. A large crew of assistants was required, and any breeze or sudden movement would have set the tube quivering uncontrollably.

Considered against this backdrop, it is easy to understand why there was a growing imperative to devise telescopes that would use mirrors rather than lenses. With a mirror, light simply bounces off the front surface and therefore does not disperse into an unwanted rainbow. But, while optical technology could produce adequate lenses by the early 1600s—albeit with the defect of chromatic aberration—it was another 60 years before opticians learned the art of making suitable mirrors. Few of those opticians would have been aware that it's actually the laws of physics that make it so much harder to create an accurate reflecting surface than the equivalent lens surface.

The accepted wisdom is that the first person to succeed in this was Isaac Newton in 1668, by dint of careful experimentation in mirror polishing together with a sound theoretical understanding of both the physics of reflective surfaces and the ideal layout for the optical components. That much is certainly true—the arrangement he proposed is still the most common form of reflecting telescope in use by amateur astronomers. It's no accident that it is known as a Newtonian telescope.

But, up in Scotland, there was a man a little older than Newton who had, before this, arrived at a telescope design of his own and had made a brave attempt to turn it into reality. This fellow was James Gregory, a gifted Aberdeenshire mathematician who had published the design in his book *The Advance of Optics* early in 1663, along with a discussion of the relative merits of telescopes using only mirrors or only lenses. Gregory's design used a combination of both but required two accurately manufactured dished mirrors in order to work properly. Of unequal diameters and curvatures, these were referred to as the 'primary' and 'secondary' mirrors of the telescope, the primary being the larger (and shallower) of the two.

Gregory was nowhere near as practical a man as Newton, so, on a trip to London late in 1662 in connection with the publication of *The Advance of Optics*, he had engaged the services of an optician called Richard Reeve and his assistant Christopher Cock to grind and polish the metal mirrors. (At that time, mirrors were made of a brittle alloy called 'speculum metal' rather than glass.) Reeve (whose name is variously spelled Rive, Reive, Rives or Reeves) was one of the most accomplished optical workers of the day and a thoroughly interesting character in his own right. Not long after his work with Gregory, he slew his wife—possibly unintentionally—but received a royal pardon on account of his optical skills. And, indeed, it seems that the only thing that stopped him producing a fully working reflecting telescope for Gregory early in 1663 was the Scotsman's haste to be off on a sabbatical tour of the Continent. This is revealed in a letter written on 23 September 1672 by Gregory (who was by then regius professor of mathematics at the University of St Andrews) to John Collins, a prominent fellow of the Royal Society.

> As for my experiment with Mr Rives, he could not polish the large concave upon the tool . . . Upon this account, & being about to go abroad; I thought it not worth the pains to trouble myself anie further with it, so that the tube was never made; yet I made some tryals both with a litle concave & convex speculum; which wer but rude, seing I had but transient views of the object.

This suggests that Gregory came close to building the world's first reflecting telescope five years before Newton did. While Reeve had not managed to polish the primary mirror to Gregory's satisfaction, it had been good enough to provide 'transient views' when tested with the secondary one.

What is even more notable from the letter is that Gregory was quite clear that he used both concave *and* convex secondary ('litle') mirrors in separate tests—that is, mirrors that are dished inwards and outwards respectively. The significance of this is that today a reflecting telescope design that uses a combination of two unequal concave mirrors is called a Gregorian telescope, but one that uses a large concave mirror with a small convex mirror—as also described by Gregory—is called a Cassegrain telescope.

Gregorian is obviously from Gregory, but where does the name Cassegrain come from? There are some Australian astronomers who are convinced that it's named after a vineyard near the mouth of the Hastings River, in New South Wales, but I suspect they are suffering delusions induced by some of that same establishment's fine products. Cassegrain was, of course, a person—a seventeenth-century mathematician who lived in France and who probably wasn't averse to a drop of sauvignon blanc himself. That much is well known, but no sooner did Cassegrain pop up in the historical record than he promptly dropped out of it again, vanishing almost without trace. In fact, even his original appearance in the annals of the telescope was at second hand, for his idea had been put forward on his behalf, in 1672, by another Frenchman, Henri de Bercé.

To understand why Cassegrain performed this astonishingly successful disappearing act, you have to look no further than Newton's caustic response to the Frenchman's invention in a letter he wrote to Henry Oldenburg, secretary of the Royal Society, on 4 May 1672. This was published in the *Philosophical Transactions* (the society's journal) and, no doubt, quickly found its way to France. Basically, it poured scorn on the idea, suggesting that

Cassegrain should try manufacturing one of the telescopes before he made such announcements.

> I could wish therefore M. Cassegraine had tryed his designe before he divulged it; But if, for further satisfaction, he please hereafter to try it, I beleive the successe will informe him, that such projects are of little moment till they be put in practise.

But, in later correspondence, it's also evident that Newton believed Gregory must have been fully appraised of the Cassegrain design when he wrote *The Advance of Optics*, almost a decade earlier. As Newton made clear in a letter to Collins dated 10 December 1672, Cassegrain's 'invention' was not his to invent—Gregory had already done it.

> I doubt not but when he [Gregory] wrote his *Optica promota* [*The Advance of Optics*] he could have described more fashions then [*sic*] one of these Telescopes & perhaps have run through all the possible cases of them if he had then thought it worth his paines. Because M. Cassegrain propounded his supposed invention pompously, as if the main business was in the contrivance of these instruments I thought fit to signify that that was none of his contrivance, nor so advantageous as he imagined.

It would be a bold individual indeed who would take on Isaac Newton, given his mushrooming reputation, and Cassegrain did what most of us would have done and shrank back into obscurity. So complete was his disappearance that it was only in 1997 that French astronomers were able to identify him as Laurent Cassegrain, who was born in Chartres, in the Eure-et-Loir département in northern France, in about 1629 and died in nearby Chaudon on 31 August 1693.

Despite these remarkable events, Cassegrain's name is still attached to the most common arrangement of mirrors found in the largest observatory telescopes. Gregory's contribution has slipped out of prominence, and that has happened with much of his work. In mathematics, Gregory was probably the equal of Newton and, under different circumstances, might have gone on to great things. But his unassuming personality meant that he was always in awe of Newton and often failed to publish his own ideas as a consequence. And, unlike Newton, who lived to the age of 84, Gregory died while still a young man, suffering a stroke when he was 36, only a year after he had taken up the chair of mathematics at the University of Edinburgh, in October 1674.

On one matter, Newton and Gregory were in complete agreement: their regard for the late Jeremiah Horrocks following the publication of his *Posthumous Works* in 1672. Each of the men wrote to John Collins about the book. Newton said, 'I am very glad that . . . the world will enjoy the writings of the excellent Astronomers Mr Horrox & Hevelius,' while Gregory was characteristically more sympathetic:

> I received . . . *Horrocci posthuma* [*Posthumous Works*], for which I must aknowledge my self exceedinglie engaged to you: I have perused him & am satisfied with him beyond measure; it was a great loss that he dyed so young; many naughtie fellows live till 80.

Ah, yes. Even today you've got to watch out for those naughty 80-year-olds.

With both Horrocks and Gregory fated to die at a young age, it was left to Newton to soar in his thinking regarding the big questions of the mechanisms of the Solar System. And, as the tea card album sketch hints, this culminated in his work on gravitation.

We know that Newton's ideas on gravity date from an enforced stay at his home in Lincolnshire in 1665, when the University of Cambridge had closed its doors to prevent an outbreak of the plague that was devastating London's population. There, he linked the downward force pulling an apple to the ground with the force that holds the Moon in its orbit. The mathematical relationship governing the nature of that force gradually took shape in his mind, and, by 1684, when he received an enthusiastic visit in Cambridge from Edmond Halley, he had formulated its exact form.

Halley facilitated the publication of this work, along with Newton's theories of the motion of objects through space and their motion through a dense medium such as water. Together, the various chapters formed *The Mathematical Principles of Natural Philosophy* (*Philosophiae naturalis principia mathematica*), published in 1687 and perhaps the most far-reaching scientific book ever written. Known today simply as the *Principia*, this extraordinary work not only explained the motions of planets and satellites as deduced by Kepler but extended the study to projectiles shot from guns, pendulums, comets, tides and the more subtle motions of the Earth and Moon. With marvellous insight, Newton's conclusions ranged from the simple idea that all heavenly bodies are in a state of mutual attraction to the unbelievable notion that artificial satellites could be made to orbit our planet. The *Principia* was earth-shattering in its influence, solving most of the problems then current in astronomy and setting the course of scientific research for the next two centuries. It was a very hard act to follow.

HALLOWED GROUND

Many of the places we have roamed through in this chapter were destinations for the Stargazer I tour in 2008. Central London was particularly memorable, not only for the Royal Society and the Monument to the Great Fire of London, but also because of an incident involving two of our party. As keen astronomers from the southern hemisphere, they were eager to use their portable telescopes to explore the sights of the northern sky and went off in search of somewhere shielded from the city's lights. One of the royal parks seemed ideal for the task—but how were they to know that certain areas are out of bounds to the public after dark? They came close to being arrested for their late-night excursion and returned shamefaced to the hotel in double-quick time.

In fairness to their enthusiasm, the incident was not too different from an unexpected visit I had one night as a youngster when I was using a borrowed brass telescope on a tripod in the garden of my home in Yorkshire. The visitor was a large policeman, who was responding to a call from a neighbour about a suspicious-looking youth fooling around with a Second World War bazooka. Fortunately, I was able to reassure him that it wasn't loaded. It's the kind of thing that happens when you get into astronomy.

The Stargazers took in Yorkshire, too, with a brief stop in Leeds, not far from where William Gascoigne had lived, and the next day we embarked on a memorable visit to Isaac Newton's Cambridge. At the Institute of Astronomy we were among friends, who happily showed us their treasures.

A particular highlight of the tour, however, was our homage to James Gregory in Scotland, with visits to both St Andrews and Edinburgh. The universities of those two ancient cities are where I was educated, so this was a trip close to my heart. At St Andrews we were privileged to receive a conducted tour by a good friend and former colleague, Andrew Collier Cameron, whose research speciality is the planets of other stars. We were able to see the north–south meridian line set into the floor of the Upper Hall of the old University Library by Gregory himself, when he used the room for astronomical observations in the early 1670s. His clock is also there, along with an iron bracket used to support his spindly refracting telescope in a south-facing window. The humble bracket had a narrow escape during the mid-2000s, when it was inadvertently consigned to a skip during renovations to the building.

At St Andrews, too, we were joined by the astronomer royal for Scotland, John Campbell Brown, of Glasgow University, who is a great supporter of science outreach activities and was happy to give us a unique talk in nearby Dunfermline, where my daughter and son-in-law owned a restaurant. His memorable after-dinner presentation was illustrated not with PowerPoint slides but with magic tricks. Now there's a science communication skill I wish I could emulate.

The visit to Edinburgh was equally memorable. The Royal Observatory houses one of the jewels of the astronomical world, in the shape of the Crawford Library, a collection of rare books and manuscripts gifted to the observatory by the 26th Earl of Crawford in 1888. It is one of the most important collections in the world, and we were amazed at the generosity of the librarian Karen Moran in allowing us to handle first editions of

Copernicus' *Revolutions*, Newton's *Principia* and the works of Tycho, Galileo, Kepler and Gregory—to name just a few. A couple of the tour participants had tears in their eyes as a result of this close encounter with some of the greatest works in the whole of science.

In Edinburgh there was a classic case of coals being taken to Newcastle, in the shape of a public lecture about the local hero, James Gregory, given by a visiting Australian scientist. I wonder who that could have been?

And then we went to Bath. Why Bath? It's a beautiful city with architecture from the Georgian era of the eighteenth century and a heritage that goes back to Roman times, most of whose historic venues are within walking distance of one another. Notable attractions include the Roman baths and the Jane Austen Centre, and we were also able to take in the United Kingdom's most famous megalithic site on the way there from London—the prehistoric grandeur of Stonehenge.

But our main reason for visiting Bath was that it was where Newton's very hard act to follow was, well, followed. From 1766 to 1782, Bath was the home of William Herschel, truly one of the brightest stars in the history of astronomy. Variously described as the 'greatest astronomer of all time', the 'greatest telescope-maker of all time', the 'father of galactic astronomy' and the 'father of infrared astronomy'—all with some justification—Herschel is a towering figure in science history. His career is all the more remarkable when one considers that it was, in fact, his second career, on which he did not embark until 1773, when he was 35 years old. Before that (and, indeed, for some years beyond), he was a professional musician.

It was in Bath where Herschel rocketed to fame, when he became the first person ever to discover a planet. The five naked-eye planets—Mercury, Venus, Mars, Jupiter and Saturn—have been known since antiquity, but, on 13 March 1781, by dint of careful observation with a homemade telescope, Herschel discovered another one. It's well known that he wanted to call it *Georgium sidus*, the Georgian Star, in honour of the king, and that could be seen as a shrewd move to curry royal favour. Indeed, it was—and it did—but it was not Herschel's fault that the seventh planet of the Solar System ended up with the mildly inappropriate name by which we know it today. That was the idea of one Johann Elert Bode, and even then he can't really be blamed, as the name sounds absolutely fine in his native German.

Uranus was discovered from the garden of Herschel's modest house in New King Street, Bath, which is today the Herschel Museum of Astronomy. It is also the place where he conceived the idea of developing monster telescopes to study the enigmatic misty patches that astronomers call 'nebulae', and, along the way, of course, to impress his monarch. Herschel recognised that ever-bigger mirrors would reveal fainter objects in ever-greater detail, and it became his unfulfilled quest to find out whether *all* nebulae were made of stars or whether some were made of something else. That quest eventually inspired the greatest reflecting telescopes of the nineteenth century. It was with these heady issues filling their minds that the Stargazer tourists made their pilgrimage to the Herschel Museum. If ever a group of enthusiasts trod upon hallowed ground, this was it.

UNSAFE PRACTICES

Despite his intellectual talents, his practical abilities and his general all-round competence, William Herschel had a few blind spots. One is in an area that we take infinitely more seriously today than he evidently did—to wit, occupational health and safety. Herschel's behaviour in this regard was a complete disaster. But this is perhaps not surprising, since he was attempting to build the largest telescopes ever conceived using structural techniques that were rudimentary even by the standards of the late eighteenth century.

His most famous instruments are known by the lengths of their tubes—the Large Twenty-Foot (6.1 metres) of 1783, and the Forty-Foot (12.2 metres), completed six years later. These telescopes were built at Herschel's later home in Slough, near Windsor. They were monumental timber structures carrying metal mirrors that, in the larger telescope at least, weighed a significant fraction of a tonne. The observing position on both was high on the rigging and very exposed. They were dangerous instruments to work with, and when the Large Twenty-Foot was blown over in a gale, in March 1784, it was only by chance that no one was injured. Herschel merely noted in his journal that, 'fortunately, it is a cloudy evening so that I shall not lose time to repair the havock that has been made'.

Herschel's sister, Caroline, an able and accomplished astronomer in her own right, was much more aware of the hazards surrounding the family business. 'I could give a pretty long list of accidents which were near proving fatal to my brother as well as myself,' she wrote late in her life. Indeed, she had first-hand experience of William's neglect when, in the darkness of a winter's night, she gashed her leg badly on an iron hook hidden in the snow. Likewise, a protruding bar on a telescope caused serious injury to an eminent visiting astronomer whom we met in Chapter 2—Giuseppe Piazzi, the discoverer of the first known asteroid.

Herschel himself experienced several heart-stopping moments, including one with his younger brother, Alexander, in 1807, when a beam supporting the 1-tonne mirror of the Forty-Foot broke as it was being removed from the telescope for its regular repolishing. Fortunately, it didn't have far to fall onto its handling carriage, but Caroline noted with evident shock that 'both my brothers had a narrow escape of being crushed to death'.

Perhaps the worst episode of all occurred 26 years earlier, however, when William was carrying out his first experiments in casting large telescope mirrors in his house in Bath. In August 1781, he tried to cast a mirror 90 centimetres in diameter for a proposed Thirty-Foot (9.1-metre) telescope. The experiment failed when a quarter of a tonne of molten metal poured with explosive violence out of a broken mould onto the stone floor of the basement workshop. Herschel and his workmen were lucky to escape with their lives, and the project was abandoned. Fortunately for astronomy, Herschel later managed to get the hang of this technique.

Perhaps it is churlish to accuse Herschel of neglecting safety standards when both his telescope making and his observing pushed back the frontiers of knowledge in ways that had never been seen before and have hardly been seen since. But it remains true that he was lucky—very lucky—that no one was killed as a result of his activities. Especially since the life most likely to have been lost was his own.

Before we move on from Herschel, there is one more story of unsafe practice on his part to be told, concerning risky behaviour of a different kind. The tale was uncovered a few years ago by the Cambridge-based Herschel specialist Michael Hoskin.

In 1757, as a nineteen-year-old musician, Herschel moved from his native Hanover to London. We are told in the history books that this was to pursue his ambitions as a composer, but it now seems there was rather more to it than that. The tale may even go some way towards explaining the zeal with which Herschel immersed himself in the culture of his adopted country, discarding the Hanoverian name of Wilhelm Friedrich to become the terribly British William Frederick.

We actually know a great deal about Herschel the musician, particularly since much of his music has survived and is today readily available in recordings by leading musicians. There's no doubt that he was a composer of great talent, and if you want to acquire insight into his cheerful disposition you can hardly do better than listen to some of his organ works, as recorded, for example, by the modern-day French astronomer-musician Dominique Proust. Indeed, participants in the Stargazer II tour, in 2010, were delighted when Proust gave us a private recital at the church of Notre-Dame de l'Assomption in Meudon, near Paris, after showing us the highlights of the Meudon Observatory. Given the optimistic nature of Herschel's music, it's surprising to discover that his departure from Hanover took place in the midst of chaotic upheaval.

He had been born into a musical family, the son of an oboe-player in the Hanoverian Guards. In 1753 he followed his father and his older brother, Jacob, into the regimental band, but within a few years military responsibilities overtook musical ones. French ambitions against Hanover resulted in the defeat of the Guards at the Battle of Hastenbeck, in July 1757. William Herschel, then eighteen, was told by his father to hotfoot it home to escape the fighting. When he arrived in Hanover, however, his mother told him he would be far better off with his regiment, since a civil defence force was being mustered in town and he risked being conscripted into it. At least in the army he was officially a non-combatant. So back he went, stealing unnoticed to his post.

It's a measure of the concern Herschel's father had for his two sons that he then plotted to spirit them across the English Channel to escape the ongoing skirmishing with the French—who, by now, had occupied Hanover. This was not so much a problem for Jacob, whose musical talents had allowed him to revert to civilian status, but William was still in the army. So, late in 1757, the two brothers arrived safely in the United Kingdom, then ruled by the Hanoverian dynasty of the royal family. Back home, however, Herschel's father was promptly arrested 'by way of enforcing the return of the Deserter', as Caroline put it in her autobiography, although it failed to have the desired effect. It was another two years before the Hanoverian troubles subsided, with the defeat of the French at the Battle of Minden, in August 1759. Jacob quickly returned home, but that option was not open to 'the Deserter', who elected to remain in the United Kingdom. It was not until 1762 that William Herschel received his discharge from the army—probably through Jacob's influence—and was able to visit his native city once again.

Herschel might not have been happy to be labelled a 'deserter', but most of us would have done exactly the same thing in such circumstances. His father—entirely understandably—was the instigator of these events, and he paid the price with his detention. It's easy to imagine that, had he not taken the course of action he did, his gifted son might have met an early end on the battlefields of Lower Saxony—and astronomy would have been immeasurably the poorer.

Perhaps the one hint of embarrassment on Herschel's part, highlighted by Michael Hoskin, is that when he wrote of the events a quarter of a century later, Herschel was rather sparing with the truth. He explained that the war had made his situation in Hanover 'very uncomfortable', but also said that 'the known encouragement given to Music in England determined me to try my fortune abroad & accordingly about the year 1759 I came to settle in this country'. Who can blame him for glossing over the details?

ULTIMATE JOURNEY?

As a study tour destination, Hanover has much to recommend it, and no doubt will figure in future science history trips. I've been there once, but it was under quite different circumstances from the relative calm of an astronomy tour. It was a mission—something that just had to be done. My brother, John, and I took our elderly Aunt Dorothy to Hanover on a brief summertime visit that, I guess, was our way of saying thank you for all the wonderful times she had given us decades before, when we were youngsters. It was a pilgrimage that had already been made by her father and sister, and we had wanted to ensure that she made it, too.

By then, though, Aunty Dot was seriously cantankerous, more than slightly incontinent and wont to chain-smoke her way through almost every waking moment. Her manner towards John and me was reminiscent of the way you'd deal with a pair of badly behaved ten-year-olds, despite the fact that we were both in our fifties and had grown-up kids of our own. But we coped with all that and tried to give her the best experience we could.

Our destination was in a place called Harenberg, on the western outskirts of Hanover, and, as we approached through open countryside along the Harenberger Meile, we could see our objective off to the right. On a gentle south-facing slope, sparkling in the sunlight after a passing shower, stood row upon row of identical white headstones. John parked the car, and we walked among them, reading the inscriptions as we went. The most sobering aspect was the age of the brave young men whose headstones they were. Most had been between 19 and 22, barely in their prime, as they took to the skies in their thousands to inflict a mortal blow on a psychopathic regime. The neat rows stood in stark contrast to the chaos that would have surrounded those Allied airmen as they fell. Gathered from all over Lower Saxony, their bodies suffered similarly horrific injuries to those of the fallen of Marston Moor, exactly 300 years before. It made me wonder what we had learned. Of course, we were looking for one name in particular. And, yes, we found it.

What would a bunch of astronomy tourists be doing in Paris in June? Pretty much the same as anyone else, for the most part—seeing the sights and soaking up the atmosphere. As they were Stargazer II tourists, of course, there had to be a bit of study thrown in, but our genial host, Dominique Proust, made light work of that. A memorable Saturday-morning visit to the Paris Observatory, in leafy boulevard Arago, in the 14th arrondissement, gave us an insider's view of the evolution of Parisian astronomy since the observatory's foundation by Louis XIV in 1667. The magnificent building, the first national scientific institution of its kind, was truly the epitome of the Enlightenment.

Then, when we took the half-hour coach drive to Meudon later in the day, we were able to visit the observatory's biggest telescope—and the largest lens telescope in Europe. Sadly, the 83-centimetre-diameter Grande Lunette (Great Telescope) of 1889 was not accessible to visitors, due to weather damage sustained during a bad storm in December 1999. Though restoration had started, the telescope was a pitiful sight in its protective plastic sheeting when we glimpsed it through an access doorway. But its situation overlooking Paris is magnificent, and the telescope's building is itself a remarkable piece of recycling. It is built on the foundations of the eighteenth-century Château Neuf du Domaine de Meudon, which burned down in 1871.

Paris is also notable as the home of Louis-Jacques-Mandé Daguerre, pioneer of photography in the 1830s. Though it needed several decades of further development, photography revolutionised astronomy in the 1880s with its ability to detect and permanently store the images of celestial objects seen faintly through the telescope. My co-host on the trip was the great modern-day astrophotographer David Malin, and we relished his unique insight in a public lecture at the observatory.

So, our visit to Paris took in all of that and more, including one of those extraordinary coincidences that happen from time to time, when we ran into a good friend of mine from Oxford—a former colleague—on the second level of the Eiffel Tower. And the tour was capped off on its final evening with dinner on the first level, with another distinguished special guest, Ian Corbett, general secretary of the IAU. (Remember the IAU? Yes, the organisation that famously demoted Pluto is based in Paris.)

CELESTIAL SIGNATURES

But now let me tell you about someone we *didn't* celebrate in Paris. Not intentionally, in fact—I would certainly have mentioned him if I'd remembered to. Auguste Comte was a great French philosopher who lived in the city on and off during the first half of the nineteenth century. He is famous as the founder of the doctrine known as 'positivism', in which reason and scientific testing are proposed as the only sensible routes towards understanding both physical and human processes. Actually, positivism encompasses much more than that, but its basic principles are what drive all scientific enquiry. In that regard, despite his occasionally eccentric views, Comte is regarded as one of the defining figures of the philosophy of science in the nineteenth century. One of science's heroes, in fact.

There is one aspect of his work, however, that invariably brings a little smirk of satisfaction to the faces of astronomers. In one of the volumes of his *Course in Positivist Philosophy*, published in 1835, he makes some rash statements about the stars. For example, he insists that

> we shall never be able by any means to study their chemical composition . . . We shall not at all be able to determine their chemical composition or even their density . . . I regard any notion concerning the true mean temperature of the various stars as forever denied to us.

And so on. You have to have some sympathy for him, I suppose. Comte imagined that in order to work out what the stars were made of or how hot and dense they were you'd have to have physical samples from them. He believed there was no way that simply looking through telescopes at stars could ever reveal their true nature. But he was wrong.

Two years after Comte's death in 1857, the German physicist Gustav Kirchhoff made a profound discovery.

He demonstrated that every chemical element has a characteristic signature in the light it emits when burned in a flame or excited with an electric current. The hidden signature is unlocked when the light is viewed through a prism, revealing a 'barcode' of coloured lines in the rainbow spectrum that uniquely identifies the particular element.

After a further two years, in 1861, Kirchhoff succeeded in applying that knowledge on a grand scale. Working with his colleague Robert Bunsen (of burner fame) at the University of Heidelberg, he identified the chemical elements present in the Sun's atmosphere by means of corresponding lines in the rainbow spectrum of sunlight. Dark lines imprinted on the Sun's spectrum are simply the bright lines of glowing chemical elements reversed in intensity, almost like a photographic negative. This astonishing feat was proof that you don't need to have a physical sample of something to know unequivocally what it's made of. You don't even have to be anywhere near it. It was a truly remarkable discovery.

The next step was even more remarkable, however. Over in Victorian London, a 37-year-old astronomer of independent means was inspired by the work of Kirchhoff and Bunsen to wonder if the same trick could be applied to the light of other celestial objects. William Huggins had a sizeable telescope by the standards of the day—a refracting telescope 20 centimetres in diameter—and he had time on his hands. With the help of his friend William Miller, a professor of chemistry at King's College, Huggins fitted this instrument with a spectroscope—a device for examining the rainbow spectrum of light—and together they began exploring the heavens in a new way. While the Moon and planets showed essentially the same spectroscopic barcode as the Sun (not surprisingly, since they all shine with reflected sunlight), the stars took the scientists' breath away. The stars, they found, also had spectra that showed recognisable patterns. By 1864, Huggins and Miller had carefully studied 50 stars, unmistakeably identifying the barcodes they had found in their spectra as those of chemical elements found on Earth. As Huggins later wrote,

> one important object of this original spectroscopic investigation of the light of the stars and other celestial bodies, namely to discover whether the same chemical elements of those of our earth are present throughout the Universe, was most satisfactorily settled in the affirmative; a common chemistry, it was shown, exists throughout the universe.

That was an extraordinary discovery not just for scientists but for philosophers, too, blowing right out of the water Auguste Comte's ideas of things we can never know. At the same time, it gave birth to the new science of astrophysics—the physics of the stars—in which astronomers attempted to understand the processes that caused spectroscopic differences between one star and another. Today, we know that most of those differences come about because of variations in the size and temperature of the stars, but this was completely unknown territory in the 1860s.

Another property of a star is imprinted on its spectroscopic barcode: its radial velocity, or its speed towards or away from the observer. It comes about for the same reason that the siren of an emergency vehicle seems to change pitch as it goes past, a phenomenon called the Doppler effect. Light behaves rather like sound in that its 'pitch' becomes different when its source is moving. If for 'pitch' you read 'colour', you can perhaps understand how the position of the tell-tale barcode in the spectrum of a star changes slightly with its velocity, giving astronomers an unexpected celestial speedometer. Huggins made the first attempts at such delicate measurements in 1868, but it was not until twenty years later that the first reliable velocities were measured photographically, by Hermann Karl Vogel at the Potsdam Observatory.

ABSOLUTELY NEBULOUS

In his wide-ranging explorations of the spectroscopic sky, William Huggins solved one other problem that had long dogged astronomers. It concerned the true nature of nebulae—the mysterious fuzzy patches in the sky that were neither stars nor planets. As we saw in the preceding chapter, William Herschel had long wondered about these, and many astronomers believed that they were gigantic aggregations of stars that were just too far away to be separated into individual points of light. Indeed, that's what many of them were eventually discovered to be. But not all.

In 1864, Huggins turned his spectroscope onto one of these nebulae and was astonished to see the unmistakeable barcode of a glowing gas: a series of narrow, coloured lines spaced out from one another and quite different from the continuous rainbow ribbon of light crossed by dark lines that is characteristic of a star like the Sun. As he later wrote, 'the riddle of the nebulae was solved. The answer, which had come to us in the light itself, read: Not an aggregation of stars, but a luminous gas.' This triumph of scientific endeavour put astrophysics—and Huggins—firmly on the map. He must have imagined there was nothing his marvellous technique couldn't do.

Indeed, his career flourished, and eventually, in 1897, he was awarded a knighthood.

There was just one niggling problem that took a little of the shine off Huggins' discovery. Embarrassingly, most of the bright, coloured lines, or emission lines, in the spectra of nebulae could not be identified: they didn't seem to correspond with any of the chemical elements found on Earth. This was in stark contrast to the dark lines that Huggins and Miller had found in the spectra of stars. True, there were lines in the blue and violet part of the spectra of nebulae that seemed to coincide with lines known to be emitted by glowing hydrogen, but others didn't fit any known pattern. In particular, the brightest and most prominent lines, which were in the green part of the spectra, completely defied identification. This mystifying lack of correspondence was not something that could be put down to a velocity shift caused by the Doppler effect, for example; the barcodes of the nebulae simply had no terrestrial counterparts.

It was several years before further progress was made. But what happened next was perfectly logical and followed a similar conundrum that had emerged in August 1868 during a total eclipse of the Sun. On that occasion, a number of well-known scientists, including a French astronomer, Georges-Antoine-Pons Rayet, had made spectroscopic observations of solar prominences—huge clouds of glowing gas billowing from the Sun's surface. Rayet had found no fewer than nine bright emission lines, among which was one he took to be something called 'sodium D', the well-known orange-yellow emission line that gives today's sodium street lamps their characteristic colour. But further investigation by a London-based scientist, Norman (later Sir Norman) Lockyer, and his colleague, Edward (later Sir Edward) Frankland, soon revealed that this was in fact a different spectrum line—and one that did not correspond to any known substance. They therefore assumed that it originated in a chemical element unknown on Earth, which they eagerly christened 'helium', a beautiful name derived from the Sun's personification in Greek mythology, Helios. This rash act of faith in the power of spectroscopy was vindicated in 1895, when helium was extracted from the mineral cleveite by an enterprising Scottish chemist called William (later, of course, Sir William) Ramsay, who boiled it in weak sulphuric acid. It is easy to imagine the glee with which Lockyer and Frankland must have greeted the news of this feat of latter-day alchemy. For the first time, an element had been identified in a celestial object before being discovered on Earth.

Astronomers who were worried about the failure to identify the spectrum lines in nebulae took heart from the work of Lockyer and Frankland. Following their lead, they declared that the mysterious green lines in their nebulae spectra must come from an undiscovered element, which they called 'nebulium'. (Another nice name, you have to admit, albeit a little less catchy than 'helium'.) In the gung-ho climate of the day, it was the only sensible thing to do and was heralded as a further triumph of astrophysics. Problem solved.

However, as time went by, more and more experiments failed to reveal any trace of these nebulium lines in laboratory spectra, and gloom gradually settled once more over the astronomers' camp. Moreover, a new tool for chemical investigation came into vogue, in the shape of the periodic table of the elements, and this didn't seem to have any gaps among the lighter elements where you might expect nebulium to appear. Could nebulium perhaps be something that was extraordinarily rare in the

Universe? Absolutely not—the stuff was everywhere. The brilliance of the green lines testified to its abundance in the gaseous nebulae. As the twentieth century dawned, flourished and was all but obliterated in the carnage of the First World War, the mystery of nebulium deepened into a constant irritation in the minds of the world's physicists.

It was not until 1927 that the first glimmer of light began to shine on the problem—if you'll pardon the pun. In a book that was still a standard text when I was a lad (albeit, by then, in a much later edition), a gifted US astronomer threw out an illuminating suggestion. Henry Norris Russell was one of the leading lights of the Princeton University Observatory when he collaborated with his colleagues Raymond Smith Dugan and John Quincy Stewart to write a textbook entitled simply *Astronomy*. (A good title—nothing beats telling it like it is.)

In the book, Russell speculated sagely on the origin of the nebulium lines: 'It is now practically certain that they must be due not to atoms of unknown kinds but to atoms of known kinds shining under unfamiliar conditions.' He went further, suggesting that those unfamiliar conditions might be what you would find in a gas of very low density—almost akin to a vacuum. Russell knew that the coloured emission lines characteristic of an element had their origin in atoms of that particular element changing their energy levels, from more to less excited states. The excitement might come from electrical currents in the laboratory or radiation from stars in space. He then postulated that the reason for the emission of the unidentified lines could be that it took 'a relatively long time (as atomic events go) for an atom to get into the right state to emit them'—a state that, under normal laboratory conditions, would be rudely interrupted by a collision with a neighbouring atom. In the incredibly low pressure prevailing in a gaseous nebula, atomic collisions would be rare, and who knew what might happen to the states of the relatively undisturbed atoms?

The person who answered that question in characteristically brilliant and lucid fashion was another American, Ira Sprague Bowen. Working at the California Institute of Technology, Bowen had been calculating the possible energy states that could be taken on by various atoms and the colours of light that would be emitted when the atoms jumped from one energy state to another. Such transitions are, in fact, the origin of all coloured emission lines observed in the spectra of glowing gases, whether in a flame in the laboratory or in a nebula in the depths of space. Physicists had learned that the energy jumps were governed by certain selection rules and that some were permitted while others were, well, forbidden. That rather stern description was, in fact, a bit misleading, as the 'forbidden' energy jumps are not *really* forbidden. More accurately, they are highly improbable under laboratory conditions, because atoms are packed relatively closely together and bump each other into new energy states long before the forbidden radiation can be emitted. That is very different from the almost-perfect vacuum of space—even in the centre of a glowing nebula. Bowen had been pondering these various transitions while also reading Russell, Dugan and Stewart's *Astronomy*. Suddenly, the penny dropped. As he recounted in 1968,

> one night, I went down to work and came home about nine o'clock . . . and started to undress. As I

> got about half undressed, I got to thinking about what happens if atoms get into one of those states. Are they stuck there forever? Then it occurred to me, having read this [Russell's comments on nebulium], maybe they can jump if undisturbed in a nebula, but we can't see them here [in the laboratory] because of collisions . . . Well, I quickly put a reverse on my dressing, and went down to the lab again. Since I had these levels it was very easy to take these differences and check them up . . . It was a matter of minutes to establish it . . . I worked until midnight and I knew I had the answer when I went home that night.

The answer was that the so-called 'nebulium lines' were caused by a forbidden transition between different energy states of oxygen atoms. They were 'forbidden lines'. Bowen's familiarity with the energy states meant that he could quickly narrow down the possibilities and then calculate accurately the colours of light that would be emitted by the forbidden transitions. They agreed exactly with where the nebulium lines appeared. The puzzle that had dogged astrophysicists for more than 60 years was solved—and the answer lay in those shady-sounding forbidden lines: not stanzas of off-colour poetry, but ordinary atoms behaving unusually under extraordinary conditions.

In the flurry of inspired research that followed, Bowen was also able to match other previously unidentified nebula lines with terrestrial elements and, in the process, solve most of the outstanding problems of nebular astronomy. Nebulium itself was consigned to the history books—it simply did not exist. Bowen quickly dashed off a note for the *Publications of the Astronomical Society of the Pacific* and then, in 1928, presented his results in a seminal paper in the *Astrophysical Journal*. It makes pretty exciting reading, even today.

Ira Bowen was in his late twenties when he unravelled the mystery of nebulium, and he went on to a most distinguished career. For eighteen years (1946–64), he was director of the Mount Wilson and Palomar Observatories in California, overseeing the completion of both the 5-metre Hale Telescope—for 28 years the largest in the world—and its smaller sibling on Palomar Mountain, the 1.2-metre Palomar Schmidt Telescope, which is now called the Oschin Schmidt Telescope. He died, in 1973, at the age of 75.

The saga of nebulium is one of the great detective stories of astronomy. The stuff even found its way into the popular culture of the day as an epithet for mysterious things—although there can't be too many people out there these days who go around thinking the Universe is full of nebulium. On the other hand, there do seem to be plenty of folk worrying about alien invasions, Moon-landing conspiracies and faces on Mars. Science has ever to be on its guard to keep the facts high in the public consciousness.

One of the astronomer's most effective weapons in pushing back the frontiers of ignorance is the same one that William Huggins wielded to such effect in the nineteenth century—the spectroscope, and the photographic version he later developed, the spectrograph. Things have moved on a bit, though. Today's spectrographs have a level of sophistication that stretches technology to its very limits. They are so frugal with the faint whispers of light from distant objects that almost none is wasted. They no longer use Huggins-style photography to record the spectra but have special TV cameras super-refrigerated with liquid nitrogen that are almost as efficient as the laws of physics will allow. And, as we shall see, some spectrographs have been designed in such a way that they are capable of observing not just one target at a time but hundreds. In that way, the efficiency of data collection has been improved beyond all recognition in recent years.

For its part, astrophysics is still intoxicated with the wealth of information that comes from spectroscopy (as the technique is still known, despite the advent of modern spectrographs). Even the most basic observations demonstrate that the Universe is expanding and that unseen planets orbit other stars. And spectroscopy allows us to probe the limits of our understanding—for example, in the dark matter and dark energy of the Universe, two more examples of mysterious stuff from the Cosmos that have infiltrated popular culture. We will hear more about them, too, in due course.

So there you have it. As a tool for unravelling the innermost secrets of nebulae or as a harvester of hard facts from the sky, you can't beat a good spectrograph backed up by a sound theory. Spectrographs are—in the world of astronomy, at least—absolutely fabulous.

CATHEDRAL OF REASON

Of all the places in the world where I have shown folk around, there's one that easily outstrips the others in the number of tours I've led. You probably won't be surprised to hear that it's the place where I work. In many respects, it's the southern-hemisphere equivalent of the facility I just mentioned on Palomar Mountain—the one where Ira Bowen played a starring role. Please don't be deluded into thinking there's any comparison between him and me, though. If it had been up to me, we'd still be on the lookout for nebulium.

But one thing I *can* do is take another tour around this facility now, so you can come along too. The journey takes us a few hundred kilometres inland from Australia's eastern seaboard—yes, via the vineyards, if you like—to where one of the continent's most remarkable natural regions lies. It rises abruptly out of the plains of north-western New South Wales like an ancient sentinel watching over a timeless landscape. Thirteen million years ago, this was a vigorously active shield volcano. Vast quantities of ash and lava were spewed from beneath the sandstone beds of the plains to produce a broad, 1000-metre-high cone that dominated the landscape for 100 kilometres around. Gradually, though, the hot spot in the Earth's mantle (the soft rock underlying the crust) that had given it life was left behind by the steady northward drift of the Indo-Australian continental plate. As the volcano slumbered towards extinction, more moderate forces came into play. Slowly but insistently, sunshine, rain, wind and frost began laying bare its inner workings.

Today, the effect of that dissection is clear. The volcano's foundations are exposed in a maze of domes, ridges and towering spires, while between them are deep valleys clothed in eucalypts and acacias, and populated by kangaroos, koalas and emus. Brightly coloured parrots flit through the trees, while eagles soar on thermals over the mountain slopes. From the furnaces of hell has come a tranquil and beautiful place.

In 1818 the first European explorers of inland New South Wales set eyes on this 'most stupendous range of mountains, lifting their blue heads above the horizon'. The words of their discoverer, John Oxley, were spiced with awe, and no doubt it was a similar awe that led him to honour an officer in His Majesty's Treasury by naming them Arbuthnot's Range. Fortunately, this ridiculously inappropriate name didn't last. The range had been home to the Gamilaraay people for thousands of years, and their Aboriginal name soon reasserted itself. It is startlingly apt: Warrumbungle simply means 'crooked mountains'.

In the midst of all this natural grandeur, on the summit of one of the high ridges, is a unique place where primeval wilderness and modern technology collide head-on. The mountain top is called Siding Spring—though there's not much evidence of the spring until summer storms bring waterfalls cascading into life. It's home not just to kangaroos and emus but also to a dozen or so futuristic white buildings in a weird and wonderful array of shapes. One of them looks for all the world like an oversized Portaloo, while another resembles a gigantic silver Rubik's Cube, but they all have an air of clinical functionality that seems slightly incongruous in the rugged landscape.

They are here to escape the pollution of cities—the dust, the smog and the blinding light of a sky illuminated by a million street lamps. And when darkness falls a subtle transformation takes place. Great doors slide open, and the telescopes inside peer silently into the velvet depths of the sky. Their task is simple—to satisfy a human race curious about its place in the grand scheme of things. Siding Spring Observatory is a place of exploration, not of the wilderness around it but of the greater wilderness above.

The largest of the buildings is a truly gigantic structure, much bigger than the others. It can easily be seen 60 or 70 kilometres from the mountains, particularly if the sun picks out its dazzling white shape from the blue eucalypt haze. A perfect cylinder, 37 metres in diameter and 26 high, surmounted by a dome of slightly more than half a sphere. Within the cylinder are eight floors of offices, labs and workshops. On top, enclosed by the immense void of the dome, is Australia's largest optical instrument, the Anglo-Australian Telescope (AAT). The structure wouldn't look out of place in the CBD of a major city—apart from the complete absence of windows and that shiny, bald dome. You may not be surprised to hear that I've been working in this building for so many years that I've started to look like it . . .

Though not a giant by modern standards, the AAT is still one of the most powerful of its kind in the world. It's used by astronomers from all over the planet to look into the vastness of the Universe and study events in the remote past. As is well known, in this business, distance and time are equivalent by virtue of the finite speed of light. Though that is more than 1 billion kilometres per hour, it's not fast enough to allow the instantaneous transfer of information over distances much greater than the size of the Earth. Moonlight, for example, takes 1.3 seconds to arrive, while sunlight has to bowl along for 8 minutes to get here. But starlight takes years, and that's only scratching the surface of the Cosmos. The deeper into space we look, the further back in time we are seeing. The most distant objects observed with the AAT are seen as they were billions of years ago. By comparison, the age of the mountain top is a mere heartbeat.

A place so generously endowed with natural beauty as the Warrumbungle Mountains attracts a steady stream of holiday-makers. They come to experience the flora and fauna, the spectacular scenery, the peace and quiet. If they camp in the Warrumbungle National Park, they might also experience—sometimes for the first time—a truly dark night sky. No visitor to the area can fail to notice the domes on the mountain top, particularly the insistent presence of the AAT. To a few it's an eyesore, but to most it's a monument to scientific endeavour—a great cathedral of the Age of Reason. Nearly all of them, however, make the drive to Siding Spring to have a look.

The first thing they learn, as they wind their way up the steep access road, is that the AAT dome is the focus of the World's Largest Virtual Solar System Drive, which I mentioned in Chapter 2. On the mountain road, visitors pass scale models of Earth, Venus and Mercury, surprising in their smallness compared with the bulk of the dome representing the Sun. Once they enter the building, they find themselves in a lift taking them to the observing floor of the AAT. It's one of only three lifts in the whole district. The other two are a few metres away in the same building. The visitors are finally ushered into a long room, one wall of which is windowed from end to end and faces into the dome.

'Jeez! That is bloody *big*!' The reaction is always the same. Utter astonishment. Even though the AAT no longer ranks among the largest optical (visible-light) telescopes in the world, as it did when it was built, it is still a most imposing instrument. But it's not just the telescope that elicits surprise from onlookers. Because it was built at a time when science funding was more generous than it is today, the AAT still boasts one of the biggest domes of any telescope in the world, and from the inside it looks *huge*.

Astronomers from the Mount Stromlo Observatory of the Australian National University, in Canberra, first looked at the Warrumbungle Mountains as a possible location for a major observatory. Their quest was for a new site free from the growing light pollution of their home base. In 1964, the first telescope was built on Siding Spring Mountain—a 1-metre reflecting telescope made famous by a distinguished husband-and-wife team, Bart and Priscilla Bok, who used it to carry out pioneering studies of the Milky Way. It was followed by two smaller Australian National University telescopes and, perhaps more importantly, by infrastructure such as observer accommodation, power, water and a paved road to the country town of Coonabarabran, 30 kilometres away. Like the mountains themselves, Coonabarabran has a Gamilaraay name—and it is no less apt. It means 'inquisitive person'.

When the British and Australian governments looked jointly at possible sites for a new 4-metre-class telescope in the south during the late 1960s, the existing infrastructure at Siding Spring was certainly a consideration. But of prime importance was the quality of the atmosphere there, and tests revealed that it was, indeed, a suitable location for a new, large telescope. With spectroscopic conditions—clear apart from occasional thin cloud—prevailing for 65 per cent of nights and completely clear skies for up to 50 per cent, together with reasonable conditions of atmospheric turbulence (which dictates image sharpness), Siding Spring was considered to be the best place in Australia for an optical observatory. More recent site testing elsewhere in Australia has demonstrated that this is indeed true; but on a world scale the continent lacks the geographical features necessary to produce the consistent high transparency and exquisite imaging of the world's best high-altitude observatories. Thus, the AAT is likely to remain the largest telescope on Australian soil, while Australian astronomers and their British counterparts invest in international collaborations that locate their 8-metre-class (and larger) facilities on top of Mauna Kea in Hawaii, or in the high, arid Atacama Desert of northern Chile.

The planning and construction of the AAT as a joint British–Australian project took place throughout the late 1960s and early 1970s. The dished 16-tonne primary mirror was cast in the United States from a highly stable glass-ceramic material called Cervit by Owens-Illinois Inc., in April 1969, and delivered later that year to Newcastle-upon-Tyne in the United Kingdom. There, the firm Sir Howard Grubb, Parsons & Co. Ltd began the long process of polishing the glass blank to turn it into a finished mirror under the direction of the late David Brown, one of the country's most gifted optical scientists—and my first boss. In March 1973 it was declared ready for final testing, and its superb optical quality was revealed. If you imagine the 3.9-metre-diameter mirror expanded to be the size of New South Wales, then the biggest departure from a perfect surface you could find on it would be about the height of a pencil when laid on its side. Not bad.

Eventually, having satisfied all the technical requirements, the mirror was transported to Australia, and given a hero's welcome in Coonabarabran on 5 December 1973. The low-loader on which it had been carried from the New South Wales port of (rather appropriately) Newcastle was made to perform not one but two laps of honour around the small country town. No one was under any illusions about the changes this precious lump of glass would bring to the community. It was also becoming obvious that, because the AAT would be operated completely under computer control (the first large telescope to do so), it was going to be a very fine instrument indeed, with a pointing accuracy much better than any comparable facility. The telescope finally entered service in June 1975, eight months after its formal opening by Prince Charles.

The AAT was originally designed with the idea that photographic plates would be the main detectors, both for taking direct pictures of the sky and for recording the rainbow spectra of celestial objects. For the picture taking (which is technically known as 'large-format imaging'), that was indeed what transpired, and one of the unexpected contributors to the telescope's early reputation was the profusion of remarkable astronomical images made by its photographic specialist. This was none other than David Malin, who was with us in Paris a few pages ago. When, in the late 1970s, Malin pioneered a new technique for recording celestial objects in true colour, the impact was dramatic, rocketing both the telescope and Malin himself to world fame. For spectroscopy, however, you don't need the large area offered by photographic plates, and various experimental electronic cameras found favour because of their far greater sensitivity to faint light. By the late 1980s, these had evolved into the solid-state charge-coupled devices universally used today in both professional and amateur astronomy.

Two other circumstances conspired to increase the AAT's potency as a discovery machine. The first was that the southern sky was essentially unexplored by large telescopes. Even such obvious targets as the centre of the Milky Way Galaxy, our nearest neighbour galaxies (the two Magellanic Clouds, visible throughout Australia) and the closest large star clusters to the Sun had been observed only at low elevations by northern-hemisphere instruments. The second was that the challenge presented by this virgin territory had been met by the British when they had decided, in 1970, to go ahead with the construction of a wide-angle photographic survey instrument, the 1.2-metre United Kingdom Schmidt Telescope (UKST). That instrument, located a few hundred metres from the AAT at Siding Spring, was formally opened on 17 August 1973 and entered service two weeks later. Its initial task was to photograph the whole of the southern sky not covered by its near twin on Palomar Mountain during the 1950s, a job that eventually took the better part of a decade.

Many reputations were made during that frenzied period when astronomers at the UKST collaborated with their counterparts at the AAT to scout out the most interesting southern objects and follow them up with the larger telescope. It was, in many respects, a southern-hemisphere re-enactment of the success of Ira Bowen's telescopes on Mount Palomar. Eventually, in June 1988, the symbiotic relationship was formalised when the UKST became part of AAT's parent institution, the Anglo-Australian Observatory (AAO), instead of a distant outstation of the Royal Observatory, Edinburgh.

From the beginning, astronomers and engineers at the AAO showed themselves to be adept at building novel instruments for use with the telescope. For example, observations at infrared (redder-than-red) wavelengths became possible in 1979 with the introduction of the infrared photometer-spectrometer, unflatteringly known by the acronym IRPS. At a stroke, the AAO had both opened a new window on the Universe and set a trend for daft acronyms that continues to this day. But, with IRPS, the AAT could now see through dust clouds and study the earliest stages of the formation of stars in their dusty cocoons. Today's flagship instrument for infrared observation is IRIS2, a hybrid imager-spectrograph that was completed in 2003 and allows front-line astronomical research to continue through the Full Moon period, when the sky is too bright to observe faint objects in visible light. Investigations of subjects as diverse as the characteristics of Venus' upper atmosphere, weather on brown dwarf stars and filaments of galaxies in clusters—as well as survey-type work—are IRIS2's stock-in-trade.

Another new technique, while not invented at the AAO, was certainly perfected there—and has since swept through astronomy as nothing less than a revolution. It came about because of the development, in the early 1970s, of optical fibres for telecommunications—strands of glassy material a fraction of a millimetre in diameter with the extraordinary property that if you beam light in at one end it comes out almost undiminished at the other. A few years later, a handful of excited astronomers realised that these flexible, highly efficient light guides could have a remarkable application in astronomy. The idea was that you could use them to play God—in a manner of speaking—by rearranging the random distribution of stars and galaxies in the field of view of your telescope into a much neater, tidier format. To be more precise, you could allocate one fibre to each target object and then use the flexibility of the fibres to bring them all into a straight line at the other end, with each fibre emitting the light of its own target. The straight-line configuration was exactly what was needed for feeding light into a spectrograph. Thus, you could take a bundle of fibres, line up one on each target and then obtain the rainbow spectra of all your targets simultaneously. Before that, the only way to get the spectra of many objects had been to do them one at a time. The potential benefits in the rate at which data could be gathered were truly enormous, and this promised to transform our understanding of the Universe. It was the dawn of a new era in astronomy—the age of multi-object spectroscopy.

In the early 1980s a nucleus of that excited bunch of astronomers was at the AAO, with a couple more at the UKST (still, at that time, a Scottish outstation). One of those was a much-missed former astronomer-in-charge at the Schmidt Telescope, the late John Dawe, and the other was me. We kicked off fibre development there, and I took it forwards as my research project.

Working closely with the AAT's fibre optics specialist Peter Gray, I spent the 1980s developing a succession of fibre-optic instruments for the UKST. While Gray's own creations were well engineered and highly productive systems (his 'fibre-optically coupled aperture plate' revolutionised spectroscopy on the AAT), mine were string and sealing wax experiments with, I'm afraid, steadily more ridiculous names. The prototype was the 'fibre-linked array image reformatter'—FLAIR—although David Malin tended to refer to it (rather unkindly, I thought) as Watson's Folly. It did show promise, however, so in 1988 we built an improved model called the 'panoramic area coverage with higher efficiency' version, which became known as PANACHE. What else? When further improvements were suggested, I thought FINESSE might be rather an appropriate name, but another colleague put me right by suggesting that it would have to mean 'fails to interest nearly everyone save spectrograph engineers'. In the end, duly chastened, we just called it FLAIR II.

The drawback of the early fibre optics systems on both the AAT and the UKST was that they relied on each fibre being manually positioned in exact alignment with its target object. Not only was that a time-consuming process, but, in the case of Gray's system on the AAT, it had financial consequences, with the need for expensive brass plates accurately drilled with holes into which the fibres were plugged (or 'stuffed', to use the technical term) before the astronomer could observe each set of targets. Some sort of robotic positioning of the fibres was clearly necessary, and, during the 1980s, in collaboration with the University of Durham, in the United Kingdom, engineers at the AAT began experimenting with intelligent machines called 'pick-place robots' that would pick up, move and put down tiny magnets on a large steel plate. With a fibre attached to each magnet and a right-angled prism stuck on the end of each fibre, the stage was set for a transformation in the way multi-fibre spectroscopy was carried out.

That early work culminated in what are still today's workhorse fibre-positioning robots on the AAO's two telescopes. The 2-degree field, or 2dF, system (named after the diameter of the area of sky it can see) was commissioned on the AAT during the mid-1990s, while the 6dF (yes, you've guessed it—the 6-degree field) system on the UKST first saw light in 2001. Although the constructional details of these machines could hardly be more different, they both do essentially the same job. First, they take the positions of their target objects from catalogues stored in their electronic memories. Then, they place a sequence of 0.1-millimetre-diameter optical fibres, equipped with magnets and right-angled prisms, onto a metal plate in the correct position for each target (to an accuracy of 0.01 millimetres), taking no more than a few seconds per fibre—and they do it entirely automatically. Since the fibres themselves are rather delicate, another important requirement for the robots is that they try to remember not to break them.

Watching these two machines diligently going about their business is always impressive for visitors, rivalling the size of the telescope and dome in its wow factor. But some of their inherent reliability comes from the work of one particular scientist formerly at the AAO, a 'robot whisperer' of great skill called Ian Lewis. We have met him already in this chapter, completely by chance—halfway up the Eiffel Tower.

MOST PRODUCTIVE IN THE WORLD

When the AAT celebrated its 25th birthday, in 1999, it was with the recognition that the new millennium would bring challenges to a telescope that was starting to look small by world standards. More than a dozen ground-based telescopes with mirrors bigger than 6.5 metres in diameter were under construction or planned, and the Hubble Space Telescope had been producing breathtaking razor-sharp colour images of the Universe for almost a decade. Moreover, even then, astronomers had their sights on a new generation of what are called 'extremely large telescopes', with mirrors bigger than 20 metres in diameter.

The AAO had a proven record in building innovative instrumentation and already had an external projects group to make its expertise available on a commercial basis in collaboration with other Australian or British institutions. Recognising this expertise, and grabbing the scientific niches in which a 4-metre-class telescope on a less-than-perfect observing site could flourish, the observatory's management charted a future for the AAT that would keep it more than competitive in an era of larger telescopes. Specialisation in a small number of world-class auxiliary instruments rather than a large collection of hard-to-maintain bits of equipment was a key ingredient, together with an emphasis on survey astronomy—the gathering of census-style data on large populations of celestial objects.

The flagship instrument then newly commissioned—2dF—was unique on a 4-metre telescope. Its first task was a survey of the three-dimensional positions of galaxies to provide a detailed cross-section of the local Universe. By measuring the amount by which a galaxy's light is shifted to the red end of the spectrum, or its redshift, the galaxy's distance can be determined by virtue of the redshift–distance relationship discovered by Edwin Powell Hubble in 1929 (about which we'll learn more in Chapter 10). Technically, this comes about not because of the Doppler effect we met earlier but because of the expansion of the space through which the light has travelled since it was emitted. As space expands, light waves are stretched, and the light gets redder. The project was therefore known as the 2dF Galaxy Redshift Survey. It was completed in 2002 and catalogued 220 000 galaxies, resulting in a deluge of scientific papers.

In contrast with 2dF's ability to look deep into space in narrow pencil beams, the strength of 6dF on the UKST lies in its ability to perform surveys over the entire southern sky because of the telescope's wide angle of view. Thus, its first major task was a survey of 136 000 galaxies, in the 6dF Galaxy Survey, completed in 2005. That, too, has produced some spectacular advances in our understanding of the local Universe.

Today, 2dF has metamorphosed into AAOmega. (I'm afraid the name is another in-joke among optical engineers, the A-Omega product being a measure of efficiency in an instrument.) Like 2dF, AAOmega utilises 400 optical fibres allocated one to each object but feeds them to a new highly efficient and stable spectrograph. Since its completion, in 2006, AAOmega has been used for surveys of distant galaxies and also for surveys of stars in our own Milky Way Galaxy. AAOmega will remain the world's most powerful spectroscopic survey instrument for a few years to come, but, as we will see in the next chapter, a further metamorphosis is on its way.

Another string to the AAT's bow is the University College London Echelle Spectrograph (UCLES). Like IRIS2, it is a bright-time instrument—it can be used when the Moon is lighting up the sky. AAO engineers are always at pains to point out that, despite some early teething troubles, UCLES is not pronounced 'useless' but rhymes with 'chuckles'. UCLES is a survey instrument, too, but it's used to obtain extremely detailed spectra of stars one at a time. It is perhaps most famous for its contribution to our knowledge of planets in orbit around other stars. The Anglo-Australian Planet Search program has discovered around 8 per cent of all known extrasolar planets (currently about 800) by means of the Doppler wobble technique, which looks for stars being pulled to one side or the other by the gravitational attraction of their planets. UCLES has also been used for pioneering work in asteroseismology, in which minute oscillations in the surfaces of stars reveal details of their structure and age.

As well as its nightly exploration of the Universe, the AAT is frequently used for developmental work in instrument science, with recent highlights centring on the relatively new field of astrophotonics. The idea here is that once the light from your star or galaxy is inside an optical fibre, there are other clever things you can do with it rather than simply letting it out again at the other end. This manipulation of photons—particles of light—is similar to the way electrons are manipulated in electronic circuits, hence the name 'photonics'.

In collaboration with other Australian institutions and the Leibniz Institute for Astrophysics, the AAO has tested some truly exotic and ground-breaking photonic devices like Starbugs (miniaturised robots), fibre Bragg gratings, photonic spectrographs and laser-comb calibration cells. Don't worry if you have no idea what these things are. I can only just get my own head around them—and, apparently, I invented one of them. These devices, or their successors, are expected once again to radically change the way in which astronomical instruments are built and to add wholly new capabilities.

All the staff of the AAO, whether they be astronomers, instrument scientists, engineers, technicians or administrators, contribute to its functioning, and it is largely due to them that the institution has maintained a high record of productivity over the years. Repeatedly, in studies of the effectiveness of astronomical facilities worldwide, the AAT has come out at or near the top, and a recent analysis of this kind has demonstrated that the strategies adopted have paid off in keeping the telescope at the cutting edge of astronomy. Published in 2008, the analysis shows that the AAT is the first-ranked 4-metre telescope in the world, in both productivity and impact, achieving 2.3 times as many citations as its nearest competitor. Moreover, among optical telescopes of *any* size, on the ground or in space, the AAT is ranked just fifth in productivity and impact.

This extraordinary achievement is one of the highlights in a process that has recently taken the AAO into a totally new era. In 2002, the United Kingdom became a partner in the European Southern Observatory, whose telescopes are in northern Chile, and its science agency signalled its wish to withdraw from the AAT agreement. So, on 1 July 2010, the AAO underwent the biggest change in its long history, when it metamorphosed from the Anglo-Australian Observatory into the Australian Astronomical Observatory. The AAO is now a division of the Department of Industry, Innovation, Science, Research and Tertiary Education of the Australian Commonwealth government—which was delighted to find there was no need to fund the design of a new logo.

And there was another spin-off. While the AAT itself was obliged to retain its original name, the anomaly in the observatory's name that had irritated generations of Scottish, Welsh and Northern Irish astronomers was, at last, laid to rest. The misnomer that had for so long been a thorn in the side of those who sit in Pedants' Corner (and, let's face it, that means most astronomers) finally disappeared. After 36 years of high-profile misrepresentation, the Anglo in Anglo-Australian Observatory was consigned to history. So, rather than being Anglo-Australian, the observatory's astronomers have now become 'all-Australian'.

THE ALL-AUSTRALIAN ASTRONOMER

Despite the advent of cloud cameras and web-accessible meteorology data such as weather radar and satellite imaging, all astronomers like to get a first-hand impression of the condition of the sky. Whether or not there is cloud, which way it is moving, the smell of rain on the wind, perhaps—these are all signs that are difficult to read from a windowless control room, no matter how well equipped it is. Astronomers at the AAT are no exception, and the building's high exterior walkway is a place that affords not just routine weather checks but some of the most magnificent views in the world.

Let's end our tour of Australia's premier optical astronomy facility by stepping outside for a final night-time observation. From here, the vault of the Warrumbungles' pollution-free sky is visible in all its breathtaking glory. Even when there is no moonlight, the stars of the southern hemisphere shine brightly enough to allow the handrail and open-mesh flooring of the walkway to be plainly seen without a torch. Throughout much of the year, the Milky Way arcs from horizon to horizon in a radiant band, and it's possible to visualise the flattened disc of our Galaxy encircling the sky completely, if we could but see it through the Earth's dark form. Science tells us that this glowing swathe outlines the Solar System's home in the Universe. But from the walkway of the AAT it looks for all the world like the Dreamtime river of Aboriginal legend, and it's easy to feel an affinity with the first humans who observed the sky from this place.

To some Aboriginal people, the Milky Way's dark dust clouds are the head, neck and body of an emu. To others, its two glowing companions represent an old man and an old woman sitting by a camp fire, which is the star we call Achernar. Nothing the unaided eye can see gives the remotest hint that these two Magellanic Clouds are whole galaxies of stars and that the light from the nearest of them has been on its way for 170 000 years. Aboriginal legend also has much to say about the stars themselves, as they dutifully mirror the rotation of the Earth in their nightly excursion around the sky. Stories of hunters, beautiful young women, sacred creatures and munificent spirits seem to complement science's view that they are just other suns, fellow travellers in the Galaxy with our own.

For more than 10 000 years, thinking people have watched the Universe from Siding Spring Mountain and have been inspired by it. Today's all-Australian astronomer has much to be proud of, with an ancestry like that.

7 HOME TRUTHS
Digging up the ruins of our Galaxy

Home is where the heart is. So goes the old saying, and I guess it's still true, even in this age of unprecedented mobility. But, before you decide that this is a chapter for the stay-at-homes, have a think about this. How many of the old sages who delighted in trotting out such gems as 'Home is where the heart is' ever stopped to think about exactly what they meant by 'home'?

Well, it's where you live, of course.

Yes, but doesn't that depend on your viewpoint? For example, if you were a stray bacterium, a valiant survivor of NASA's pre-launch sterilisation stowed away aboard the *New Horizons* spacecraft en route to Pluto, wouldn't home be the Earth? From the perspective of an interplanetary traveller, for whom distance is measured in millions (or billions) of kilometres, home is definitely our planet rather than 22 Acacia Avenue or 7 Gasworks Lane. But, curiously, even this definition of home is a moveable feast. Once you start thinking about the stars, the scale changes again, and you're into light-years. A light-year, as I'm sure you'll remember, is the distance light travels in one year: 9.46 million million kilometres, as the crow flies. It's a *very* long way. And I've noticed that scientists studying stars and gas clouds within a few hundred light-years of the Sun tend to think of home as the Solar System as a whole—not even our particular planet. It seems that, for astronomers, warm and fuzzy thoughts of home can equally well be about their street, their planet, their Solar System or—taking things up a notch—their Galaxy.

Of all the definitions of home that we can imagine, the Milky Way Galaxy, in which our Solar System is embedded, must surely rank as the grandest. This gigantic agglomeration of a few hundred billion stars, plus copious helpings of gas and dust, and an embarrassing amount of something whose identity is still unknown—dark matter—represents the ultimate in terms of our home in space. And I guess the reason we identify with it so strongly is that beyond our Milky Way other galaxies are ten a penny, numbering perhaps 100 billion in the observable Universe—and who knows how many beyond our reach? It does make our Galaxy seem rather special.

So there you have it. Home is where the heart is, and home is the Milky Way Galaxy, splendid in its spiral structure and replete with stars. In that respect, we're all stay-at-homes, since there is no possibility with current technology of any of us ever leaving our Galaxy. But it doesn't mean we're not travellers. The Earth is moving through space as it orbits the Sun at a speed of 30 kilometres per second. And, if that's not enough travelling for you, the entire Solar System is whizzing around the centre of our Galaxy at nearly ten times the Earth's speed, taking about 200 million years to make a complete circuit. Now *that's* travelling.

GALACTIC PORTRAIT

If home is the Milky Way Galaxy, wouldn't it be nice to have a photo of it to hang over the mantelpiece, like some folk have aerial photos of their houses? But there's a problem. Because we are permanently embedded in the dusty disc of our Galaxy and see its grandeur only as the gossamer band of the Milky Way, we are effectively blind to its structure. You could no more take a photo of the whole of our Galaxy from Earth than you could take a photo of New South Wales from the forecourt of the Sydney Opera House.

It's only within the last century or so that we've had any clue as to what our Galaxy might look like from the outside. Until 1918, most astronomers believed the Solar System was somewhere near the centre of a flattened disc of stars, an idea that owed its origin to the work of our old friend William Herschel more than 100 years before. Herschel had based this notion on a technique he called 'gauging', in which he counted the number of stars in his telescope's field of view in different directions. He found that, to a first approximation, the Milky Way contains a similar density of stars (the number of stars per square degree) all the way round. Those of us who watch the sky from one season to the next know that the Milky Way is very patchy and varies greatly in star density over relatively small distances. However, when you look at the big picture, the density variations are relatively small, and Herschel could be forgiven for imagining that we're somewhere near the middle.

As the First World War drew to a close, a paper was published that turned this picture completely on its head. Written by a US astronomer called Harlow Shapley, it drew some startling conclusions from the way globular clusters are distributed around the sky. These mysterious objects—spheroidal aggregations of tens (or hundreds) of thousands of stars—were still poorly understood. Most of what was known about them had come, again, from William Herschel. Even the term 'globular cluster' had been coined by him, back in 1789. More importantly, though, Herschel had noticed that most of the clusters then known were in the southern-hemisphere sky, particularly around the constellation of Sagittarius, where the brightest portion of the Milky Way lies. Perhaps that had caused him a few niggling doubts about the Sun's central position in the Galaxy.

Shapley's breakthrough was in estimating the distances from our Solar System of 69 of these clusters, by measuring the brightness of standard candles he had found in them. The 'candles' were a particular kind of variable star, which changes in brightness with clockwork regularity—a phenomenon now known to be due to large-scale pulsations of the star's outer atmosphere. If the characteristic cycle time, or period, of the pulsation can be determined, then the intrinsic brightness of the star can be calculated. Comparison of that estimate with the star's measured brightness from our vantage point on Earth tells you its distance. When he made his estimates and then plotted out the distances, Shapley found that the clusters formed a roughly spherical distribution around a point in the direction of Sagittarius. He recognised that the globular clusters must congregate in a swarm, or halo, around the Galactic Centre since they could be seen above and below the starry haze defining the plane of our Galaxy. He had effectively found a way of cheating the obscuration blocking our view through the disc of the Milky Way Galaxy, thereby discovering its true centre.

From the distribution of the clusters, Shapley was able to deduce that our Galaxy is much, much bigger than Herschel had suspected and that the Solar System is about halfway between the centre of the Galaxy and its edge. That finding still holds good today, but Shapley's estimate of the actual distance to the Galactic Centre—65 000 light-years—is nearly three times greater than the modern value. That's because he was unaware that our view through the Galaxy is seriously impaired by dust in the Sun's neighbourhood, making distant objects appear dimmer than they should do. The dust also accounts for the relatively uniform appearance of the Milky Way as observed by Herschel. In fact, most of the stars we see when we look at the Milky Way are quite close by and in our local neighbourhood.

For all his remarkable insight, Shapley got one thing completely wrong. 'With the plan of the sidereal system here outlined,' he said, 'it appears unlikely that the spiral nebulae can be considered separate galaxies of stars.' Controversy over the true nature of the spiral nebulae was the hot topic of the era. Were they smallish objects within our Galaxy or gigantic systems of stars at unimaginable distances beyond? Shapley believed they were nearby, famously taking that position in a public event called the Great Debate against fellow astronomer Heber Curtis, in Washington in 1920. The final answer was not long in coming: in 1925, Edwin Hubble presented overwhelming evidence that they were distant 'island universes'—galaxies like our own. Once again, it was based on the standard candles of variable stars.

From that time on, astronomers assumed that the flattened disc of the Milky Way must be graced with the beautiful spiral arms they could see in other galaxies. It was not until the advent of radio astronomy, however, with its ability to reveal the signature of cold, dark hydrogen lurking in spiral arms, that they were finally detected and the first rudimentary maps of our Galaxy drawn. Today, our ability to map the disc of the Galaxy has been enhanced not only by modern radio telescopes and their 8-metre counterparts in the optical (visible) waveband but also by new instruments sensitive to infrared radiation, which is much less susceptible to the dimming effect of interstellar dust. Thus, orbiting infrared telescopes like the Spitzer Space Telescope are able to see at great distances the tell-tale aggregations of young stars that define our Galaxy's spiral structure.

Infrared astronomy also brought about a surprise in the 1990s. Astronomers had long known that spiral galaxies usually have a spheroidal bulge of old (yellowish) stars at their centre, and, since the Milky Way in Sagittarius is broadened above and below the Galactic Plane by similar old stars, they assumed that our own Galaxy would also have a nice symmetrical bulge if it could be seen from the outside. But infrared observations revealed that our Galaxy's bulge is actually a boxy bulge (yes, that really is the technical term for it), and boxy bulges are found in spiral galaxies that have a bar in the centre. Before your mind runs off to picture a row of beer-swilling aliens propping up the bar at the centre of a galaxy, let me explain that this, too, is a technical term, for a linear feature that occurs in the centre of many spiral galaxies. It's made up, like the bulge, of old stars, but they circulate around the galactic centre in such a way as to form an elongated structure looking a bit like a bar magnet. The spiral arms in these galaxies are typically connected to the ends of the bar.

Recent observations of the radial velocities of stars in the bulge of our own Milky Way Galaxy have confirmed the presence of a bar tilted at about 20 degrees to our line of sight. And the latest findings are that the two major spiral arms of our Galaxy (known as the Scutum-Centaurus and Perseus arms, after the constellations they pass through) do connect with the ends of the bar. This is extraordinary detective work on the part of astronomers in a number of institutions, including the universities of Cambridge, California and Wisconsin, and at last gives us the ability to hang a pretty detailed portrait of our Galaxy over the mantelpiece—even though it's not an actual photograph.

There's one more thing I should tell you about this portrait. Its subject—our Galaxy—is very big indeed by any Earthly standards. The diameter of its disc is estimated to be something in the region of 100 000 light-years, or 9×10^{17} kilometres, which, in plain English, is just shy of 1 billion billion kilometres. Like me, you probably have trouble getting to grips with such a distance. It's just an extremely big number and doesn't really mean much at all. But we can use our imaginary mantelpiece portrait of the Galaxy to get a feel for just how big it is. The picture is, perhaps, 50 centimetres across, nicely framing the graceful spiral arms of our home in the Universe. But now imagine the portrait expanded to be the size of the Earth—a planet-sized depiction of our Galaxy. On that scale, what do you think the separation of the Earth and the Sun would be? I'll tell you, before you dash off for your pocket calculator—it's 2 millimetres. Absolutely minuscule. I can't think of any more cogent indicator of the difference in scale between planetary systems and galaxies. Galaxies are *huge*.

GALAXIES OF THE IMAGINATION

When we gaze out from the confines of our Galaxy to its near neighbours and beyond, we find galaxies in a profusion of different shapes, or morphologies, to use the official term. Even if we ignore elliptical (oval) and irregular galaxies, and look at only the spirals, the variety is startling. Some have tightly wound spiral arms, some are loosely wound; some are dominated by a huge central bulge, while some have almost no bulge at all. And, as we have seen, some have a distinct bar across the centre.

In 1926, Edwin Hubble made the first attempt to categorise galaxies, producing his famous tuning fork diagram. At the base of the fork are smooth, almost spherical elliptical galaxies, whose shapes become steadily more flattened as one progresses up the handle, until the diagram divides into the two prongs of the fork, which are populated by barred and non-barred spiral galaxies respectively. At the tips of the prongs are the most loosely wound spirals in each category. Hubble depicted a gradual progression from the most symmetrical and uniform galaxies—the ellipticals—to the loosely wound, almost fragmentary forms of the extreme spirals.

Hubble thought he was seeing an evolutionary sequence in this diagram—a series of similar galaxies captured at different eras in their development. Thus, he called the ellipticals 'early-type' galaxies and the loose spirals 'late' galaxies. We now know that this interpretation is incorrect, but we still use his terminology. You have to admire Hubble's thinking: his speculation about the evolution of objects whose true nature had been only so recently discovered is a measure of his innovation in the field.

Things have moved on a great deal in our understanding of the evolution of galaxies since Hubble's time. With the advent of computers, from about the 1960s, astronomers have been able to build ever more sophisticated mathematical models of the physical processes that result in today's galaxies. They begin with instabilities in the primordial clouds of hydrogen that populated the early Universe and wind up with galaxies that look strikingly similar to those we're surrounded by—stars, dust, spiral arms and all.

In the early 2000s, the astronomers who build these models (who are known in the trade as 'theorists') found that their computers were so powerful that they could simulate galaxies almost one star at a time, making highly detailed representations. They discovered, for example, that the bars in barred spiral galaxies are probably the result of instabilities in the underlying circulation of stars. They also discovered that galaxy evolution has some rather sinister aspects, with strong hints that large galaxies grow by cannibalising their smaller neighbours. Naturally, the theorists wished to compare their models with the real thing, to determine how accurately they understood the physical processes at work in galaxy evolution. In particular, they wanted to be able to compare the details of their models with the best-observed of all galaxies—our own. And it wasn't just in our Galaxy's appearance that they were interested, but in the motions and chemistry of its constituent stars.

Sadly, the theorists were in for a disappointment. While they could build models simulating millions—even billions—of stars, the number of stars for which observational astronomers had detailed information was paltry in comparison, perhaps only 20 000 or 30 000. The data were principally radial velocities gathered over decades by analysing the barcode of information revealed when the light of a star is spread into its rainbow colours in the instrument we met in the last chapter, the spectrograph. Such velocities are essential for understanding the motion of stars in our Galaxy, but spectroscopy also has the potential to show the physical and chemical condition of a star's atmosphere, giving tell-tale pointers to its age and place of origin. Those measurements, however, had been made for an even smaller number of stars. 'Well,' said the theorists to the observers, 'you'd better get on with it, then, hadn't you?'

Fortunately, the wherewithal for 'getting on with it' was already at hand. Previous generations of observational astronomers had had no alternative but to observe one star at a time if they wanted spectroscopic details—which is why such information was so sparse. But, as we also found in the last chapter, a new technique had been developed utilising optical fibres to observe many stars simultaneously. In fact, the multi-fibre technique was originally developed to observe distant galaxies dozens at a time for studies of the wider Universe, but it could equally well be applied to stars. Indeed, one of my few claims to fame is that, back in 1982, I became the first person in the world to observe stars rather than galaxies with a multi-fibre system, using Peter Gray's pioneering fibre-optically coupled aperture plate device on the AAT.

When the 2dF instrument replaced earlier multi-fibre systems at the AAT, in the mid-1990s, it had an immediate impact on survey astronomy, in which statistics of a population census style are gathered on extremely large numbers of celestial objects. And, with the advent of this and other fibre-optic systems on the world's great telescopes, the stage was set for a revolution in our understanding of the mechanics of our Galaxy. Suddenly, we had the potential to collect spectroscopic data not on handfuls of stars but on hundreds of thousands—or even millions.

GALACTIC ARCHAEOLOGY

The idea of using observations of large numbers of stars to probe the evolutionary history of our Galaxy has come to be known as Galactic archaeology. In some ways this is a poor metaphor, since what we are looking for are fossils from past events in the Galaxy's history. Galactic palaeontology, however, doesn't have quite the same ring to it, and I suppose it's true that one of the things we are doing in Galactic archaeology is attempting to dig up the ruins of the past—often in the form of other galaxies that have been swallowed by our own.

Once again, these kinds of studies have their roots in the last century. Harlow Shapley's 1918 paper on globular clusters, for example, contained a hint that he was thinking along evolutionary lines. He speculated that perhaps the globular clusters he had observed above and below the disc of our Galaxy might, if they crossed the disc, be disrupted by gravity so that they would begin to fall apart. Specifically, he wondered whether the tightly formed globulars would end up looking like loose, open star clusters, typified by well-known star groups like the Pleiades and the Hyades (at the head of Taurus, the bull). With the benefit of almost a century of further research, we now know Shapley's speculations were not correct. The open clusters are actually composed of stars that are much younger than those in the globulars and are themselves sites of recent star formation. However, full marks to Shapley for trying.

It was already known in 1918 that certain groups of stars—for example, most of the stars in the well-known northern-hemisphere star group called the Plough—had a common velocity through space. This suggested that these moving groups were the recently dispersed remnants of open clusters, representing a later stage in their evolution. Moving groups were first observed during the nineteenth century but were studied much more thoroughly in the 1960s, by Olin Jeuck Eggen, a well-known US astronomer (and one-time director of the Mount Stromlo Observatory). In particular, Eggen noted that the best way to find moving groups is to look in what we now call 'phase space' rather than real space.

What's phase space? OK, imagine the three coordinates of distance found in normal three-dimensional space (right–left, up–down, backwards–forwards, usually denoted by x, y and z) replaced with the equivalent three components of velocity—the speed of an object along each of those x, y and z axes. It means that any point in phase space is representative of the way something is moving relative to a different point that's at rest (whose three velocity components are zero). Thus, a star's position in phase space will tell you not where it is in the Galaxy but how fast and in what direction it's moving. Formally, it is the star's velocity rather than its position.

The great thing about this is that if you can plot the positions of a large number of stars in phase space, those that group together are, by definition, moving with similar velocities—that is, at a similar speed and in a similar direction. They therefore constitute a moving group, irrespective of whether they cluster together in the same part of the sky. This gives astronomers the capability of identifying groups of stars that are moving together through the Sun's neighbourhood even though there is no obvious connection between them. In fact, the stars can be on opposite sides of the sky and still be members of the same moving group if they occupy the same region of phase space.

Further confirmation of the connection between stars comes from other indicators that can be gleaned from their spectra. Some features in the barcode of information are peculiar to each individual star, such as indicators of temperature and surface gravity (which is related to size). However, there are other parameters that will be common to all members of the group, and these can also be deduced from the spectra. One is age, since stars are believed to form in clusters at more or less the same time. The age of a star is actually rather difficult to measure, but help is at hand from a uniquely astronomical eccentricity called 'metallicity'. This refers to the amount of metal in a star's atmosphere compared with the amount of hydrogen—its relative abundance, in other words. What's eccentric about that? Only that astronomers consider any element that is not hydrogen or helium to be a metal. Bizarrely, that means oxygen and nitrogen are metals, as well as iron, copper and lead. Weird. Notwithstanding its slightly comical definition, metallicity is helpful because it's the same for all stars that were formed in the same cloud of gas and dust.

The technique of spotting connections between stars by their motion and chemistry is extremely powerful. As an example, consider a well-known group of stars with velocities similar to that of the bright star Arcturus, identified as the Arcturus Moving Group by Olin Eggen. Since Eggen's time, this group has been thought to be the dissipated remnants of an old open cluster, rather like the Plough Moving Group mentioned earlier. But, more recently, its advanced age (about ten billion years), its leisurely rotational velocity about the Galactic Centre and its low metallicity suggest that it doesn't belong to the youthful population of our Galaxy's disc, which is where open clusters are found. Rather, it belongs to something now called the 'thick disc', a tenuous extension of older stars above and below the Galaxy's thin disc. So where has it come from? A recent suggestion is that in the Arcturus Moving Group we are actually seeing a stream of stars not from a local cluster formed in our Galaxy but from the remnants of a dwarf galaxy that has been gobbled up by our own. (In the trade, we use the more polite term 'accreted', but 'gobbled up' is what it means.) This ties in with the phenomenon of star streams, which are seen both in our own and in other nearby galaxies—where they sometimes appear as beautiful, veil-like structures.

According to the latest models of galaxy evolution, large galaxies begin their mature lives surrounded by scores of smaller ones—dwarf galaxies. These, however, gradually spiral inwards under the gravitational influence of their giant neighbours. As they do so, they are tidally disrupted to extend filigree trails of stars in front of and behind them. Essentially, they are forcibly spread out along their orbits as they circulate in their death throes.

The best-known star streams in our Galaxy are the Magellanic Stream, originating in our two prominent neighbour galaxies, the Large and Small Magellanic Clouds, and the Sagittarius Stream, originating in the Sagittarius Dwarf Galaxy. The Magellanic Stream was, in fact, first detected by radio astronomers by virtue of the trail of cold hydrogen accompanying its stars. Other known star streams represent the debris of various dwarf galaxies or ancient globular clusters that have succumbed to the insistent tidal attraction of our Galaxy.

We have known for many years that the tenuous spheroidal halos of galaxies are populated not only by globular clusters but by old stars, too; indeed, we can see them in other galaxies as well as our own. What has emerged only recently, however, is the idea that perhaps most of these old halo stars began their lives as members of other, dwarf galaxies circulating around their voracious companions. If that is true, then all galaxies—and ours in particular—should retain the fossilised evidence of accreted dwarf galaxies in the stars of their halos. While we are just beginning to detect such fossil streams visually in our own and other, nearby galaxies, the potential of Galactic archaeology—with its large-scale star velocity surveys—is to reveal many more.

With perfect timing, just as I was writing these words, my colleague, Mary Williams, of the Leibniz Institute for Astrophysics, was busy discovering exactly one such star stream. Like me, Williams is a Raver—one of a group of 60 or so scientists from nine countries who are involved with a half-million star survey called the Radial Velocity Experiment (RAVE). The experiment is being carried out on the AAO's 1.2-metre UK Schmidt Telescope, equipped with the 6dF multi-fibre system we met in the last chapter. The survey has occupied all the available time on the UKST since the completion of the 6dF Galaxy Survey, in 2005.

Williams' new discovery is called the Aquarius Stream, a group of stars identified as special from the 400 000 stars whose speeds have so far been measured by RAVE. Unable to resist the punning potential, she entitled her paper in the *Astrophysical Journal* 'The Dawning of the Stream of Aquarius in RAVE'. She went on to announce her discovery at an international astronomy conference with the immortal words 'I have a stream . . .'

The newly discovered stream gets its name from the constellation of Aquarius, where most of the stars are located. Tracking back from the speeds measured today, it appears that the Aquarius Stream represents the remains of a dwarf galaxy that was absorbed into our Milky Way Galaxy 700 million years ago. The discovery is particularly striking since most of the stars lie close to the disc of the Milky Way itself, making it harder to distinguish them from the general background. As well as being an astronomer who delights in punning her way through the Universe, Williams has a very keen eye for patterns of velocities in phase space.

An important component of the halos of galaxies that I have barely mentioned also falls within the province of the Galactic archaeologist. We know that galaxies must be embedded in spheroidal blobs of something massive but invisible for their discs to remain intact as they spin. This mysterious stuff—the enigmatic dark matter—is thought to outweigh normal matter by about five to one, and to have played a vital role in the formation of galaxies in the early Universe. We will meet dark matter again in Chapter 10, but for now I will simply note that Galactic archaeologists can play a major part in the hunt to discover what it is, since their stars are tugged not only by the gravitation of visible objects but by dark matter, too.

So what are the hot topics in Galactic archaeology? Since this book has a subplot about astronomers' travels, I think we are at liberty to do a bit of virtual eavesdropping on a recent international workshop. OK, I'll admit that I'm particularly fond of this one, since I chaired its local organising committee, but it is typical of the way astronomers get together to discuss current issues in their science—sometimes quite heatedly. It took place in 2009 in an iconic (some might say clichéd) location: Palm Cove, close to the Great Barrier Reef, in north-eastern Australia.

If you're going to hold a workshop in a clichéd location, you might as well give it a clichéd title, too, so we called it Overcoming Great Barriers in Galactic Archaeology. (Groan.) But I'm glad to say that the meeting itself threw clichés out of the window and got on with some serious business in understanding the archaeology of our own Galaxy and its near neighbours in the Local Group of galaxies.

Organised by the AAO in association with the Leibniz Institute of Astrophysics, the Australian National University and the University of Sydney, the aim of the workshop was to provide a forum in which our current knowledge could be assessed and future strategies mapped out for both observers and theorists. It was attended by 40 of the world's leading practitioners in the field. In keeping with the tranquillity of Palm Cove, its format was relaxed—which was just as well, since it rained every day. However, the workshop unashamedly reverted to cliché on its final day, with a hugely popular cruise to one of the sunny islands of the reef, Michaelmas Cay.

Everyone agreed that the timing of the workshop was most appropriate. Two of the ongoing spectroscopic megasurveys that had been made possible by fibre optics technology had reached significant totals in the number of stars for which they had gathered data. These were RAVE on the UKST and the Sloan Extension for Galactic Understanding and Exploration, or SEGUE, which utilises the 2.5-metre Sloan Telescope, in New Mexico. Both surveys had accumulated datasets of around half a million spectra and were beginning to produce significant findings in Galactic archaeology.

The workshop also heard about surveys still on the horizon. New ground-based instruments such as the Australian National University's SkyMapper telescope at Siding Spring Observatory and an unusual telescope in China called the Large Sky Area Multi-Object Fiber

Baffling in its Expressionist modernity, the Einstein Tower was built at Potsdam shortly after the First World War to verify general relativity with observations of the Sun. (*Fred Watson.*)

The fastest spacecraft ever launched, *New Horizons* lifts off from the Kennedy Space Center at the start of its long voyage to Pluto in January 2006. Rendezvous will be in July 2015. (*NASA.*)

The die is cast. Delegates at the 2006 General Assembly of the International Astronomical Union vote on the definition of a planet—and relegate Pluto to dwarf-planet status. (*IAU/Robert Hurt (SSC)*.)

Looking decidedly unwell, a 3300-year-old participant in ritual warfare decorates the mural wall at Sechín, near Chankillo. Note his long thumbnails, which are handy for gouging out the eyes of enemies. (*Fred Watson*.)

Archaeologist Iván Ghezzi stands near the hilltop temple-fortress at Chankillo, with the mysterious Thirteen Towers running along the ridge almost a kilometre behind him. (*Fred Watson.*)

Stargazers ancient and modern in the city of Prague. (*Unknown.*)

Galileo fails to impress the muses of the sciences in the frontispiece of an early edition of his collected works. Although he did not invent the telescope, he used a refined version to make the discoveries symbolically portrayed in the sky. (*Crawford Library, Edinburgh/Fred Watson.*)

The humble back yard in Bath from where the planet Uranus was discovered by William Herschel on 13 March 1781. He wanted to call it George but, sadly, was overruled. (*Fred Watson.*)

Looming behind the Siding Spring Observatory Lodge, the giant bulk of the Anglo-Australian Telescope dome is readied for a night's work. (*Ben Wrigley.*)

High-tech in the bush. Astronomer Rob Sharp peers into one of the two cameras of the AAOmega spectrograph in its enclosure, deep inside the Anglo-Australian Telescope dome. (*Barnaby Norris.*)

A spectacular barred spiral galaxy, NGC 1300, captured by the Hubble Space Telescope. The galaxy's central bar is much more prominent than our own Galaxy's would be if we could see it from the outside. (*Hubble Heritage Team; ESA; NASA.*)

Some 40 million light-years distant, the edge-on galaxy NGC 5907 shows evidence of galactic cannibalism in the faint streams of stars that encircle it. (*R. Jay Gabany, Blackbird Observatory, via NASA APOD.*)

In this telephoto image NASA's *Curiosity* rover gazes towards its destination, 16 kilometres away. The layered rocks of Mount Sharp are expected to reveal much about Mars' geological history. (*NASA/JPL-Caltech/MSSS.*)

Queen Elizabeth confronts the Universe in a special NASA demonstration of the Cosmic Microwave Background Radiation. The speckled pattern is due to temperature variations in the flash of the Big Bang, while the encircling red glow is foreground radiation from our own Galaxy. (*NASA/GSFC.*)

A youthful Fred Garnett during his flying training in 1941. He was nineteen when this photo was taken. (*Unknown.*)

Looking like a giant celestial butterfly hovering near the Moon, an auroral corona bursts over the skies of Lyngenfjord in the Norwegian Arctic. (*Anne Spencer.*)

Spectroscopic Telescope, or simply the Guoshoujing Telescope, will have a major impact on Galactic archaeology. Perhaps an even bigger impression, however, will be made by a new instrument being built for the AAT. This is the High Efficiency and Resolution Multi-Element Spectrograph (HERMES), the third incarnation of 2dF, with its 400-fibre robotic positioner. HERMES differs from its predecessor, AAOmega, in that it has been designed especially for Galactic archaeology. It is what is known as a 'high-dispersion instrument', which means that the spectrum is spread out much more than in other spectrographs, allowing finer detail to be seen in the barcode. Of course, that dilutes the light to a greater extent, meaning that HERMES has to be extremely frugal with the starlight it receives. In technical terms, it means the spectrograph has to have high optical efficiency.

The quest for finer detail in the spectrum is driven by the need to explore the shape of the spectrum lines originating in the chemical elements in stars' atmospheres. Nineteen particular elements have been targeted, ranging from the commonplace (such as oxygen and calcium) to the exotic (like yttrium)—as well as hydrogen and helium. They have been chosen for their value in diagnosing stars that might have a common origin, in a technique known as 'chemical tagging' (a term coined by Australian astronomers Ken Freeman and Joss Hawthorn). But because the lines originating in these elements are widely separated along the rainbow spectrum of each star, and it would be impossible to observe the whole spectrum in fine detail, only selected areas are targeted. For this reason, HERMES has four separate cameras, each looking at one small region of the spectrum. It is what practitioners refer to as a 'four-beam spectrograph'. Gosh, who thinks of these innovative names?

With 2013 as the target date for starting operations, HERMES is set to provide insights into the chemical history of our Galaxy in a survey known as Galactic Archaeology with HERMES (GALAH). That peculiarly Australian mixed metaphor is brought to life on the project's website, where a classical Hermes sprints through the southern Milky Way holding a fluttering galah in his hand . . . The information from GALAH will dovetail extremely well with that from another project on the horizon. This is in a different financial league, however, being a spacecraft that will be launched by the European Space Agency at about the same time as HERMES comes on stream. Called Gaia, the project is especially ambitious and is expected to yield the positions, distances and motions of up to a billion stars with unprecedented precision. At last, we can see the prospect of hard information about our Galaxy rivalling the theorists' models.

We clearly stand at a significant moment in the development of observational Galactic astronomy. It rivals the recent achievements in precision cosmology, the study of the history and evolution of the Universe as a whole, which have resulted from large-scale galaxy surveys and the detailed exploration of the remnant flash of the Big Bang, or cosmic microwave background radiation. We will explore these topics further in our travels in Chapter 10, but, for now, back to Galactic archaeology and the great barriers that still need to be overcome.

Where do these barriers lie? Principally, they are at the interface between our best theoretical understanding of the formation and evolution of galaxies and what we observe. While it is true that, in broad terms, 'standard' theoretical models will produce galaxies a lot like our own, the devil is in the detail. And, in contrast to the situation prior to the start of the RAVE survey, when, as we have seen, radial velocities and physical parameters were known for only 20 000 or 30 000 stars in our Galaxy and a mere handful beyond, we are now data-rich—and getting richer—in this field. Thus, the challenge is being passed back to the theorists, who are now having to build galaxies using much more sophisticated models. They are called 'chemo-hydrodynamical' models (and you thought Eyjafjallajökull was hard to get your tongue around), and they incorporate not only the motions of stars and gas but also the evolution of their chemical constituents.

In recent years, our understanding of the overall structure of our Galaxy has gone through a revolution. We used to think there were three basic components—the disc, the bulge and the halo. But now the picture has fragmented into thin and thick disc populations and inner and outer halo components that have differing systematic motions. And some aspects of the bulge don't fit into any straightforward picture. Complexity is clearly the name of the game in today's view of the Galaxy—but this should be no surprise, since it's exactly what we see in other galaxies.

From the overall properties of our Galaxy, the workshop at Palm Cove turned its attention to the individual stars and star clusters contained within it, and here the great barriers seemed to get significantly more reef-like. Questions ranged from whether we properly understand the chemical composition of the best studied of all stars—the Sun—to the failure of calculations of the by-products of the Big Bang to reproduce the observed amount of a particular type of lithium in ancient halo dwarf stars. Few clues seem to be available to solve this lithium problem, although some very recent work in the United States now hints at a solution involving the elusive dark matter.

Another intriguing question concerned a class of almost-invisible dwarf galaxies that have been discovered by clever analysis techniques around both our own Galaxy and the Andromeda Galaxy—the nearest comparable galaxy to our own. Known as 'ultra-faint dwarf spheroidal galaxies', they are suspected by Galactic archaeologists to have fed stars into the outer halo of our Galaxy. Since their metallicity signatures closely match those of stars found in the outer halo, this seems likely. In fact, some of the stars in the dwarf spheroidal galaxies are very metal-poor indeed—they are virtually unpolluted by any chemical elements heavier than hydrogen and helium. It suggests their ancestry dates back to the very early Universe, before subsequent generations of stars enriched the gas between the stars with heavier elements.

The potential of the technique we met earlier—chemical tagging—to reveal common origins in disparate groups of stars in our Galaxy also found favour in discussions of stellar streams in the solar neighbourhood. With the new datasets, much more structure in phase space is revealed than was possible in Olin Eggen's time, suggesting the existence of multiple streams where previously only single ones had been identified. How many of these streams really are the debris of gobbled-up globular clusters and dwarf spheroidal galaxies? Could some of them be due instead to resonances akin to an echo in a room, but in this case linking the rotation of the stars about the Galactic Centre, and the rotation of the Galactic Bar and Disc? Chemical tagging offers the best way to separate these possibilities.

Thus, the detailed measurement of the physical properties of very large numbers of stars is essential to help disentangle such problems. Metallicity, surface gravity, effective temperature and so on are all required for us to understand the origin of individual stars. This will allow us to build up a population census of the Sun's neighbourhood that is entirely unprecedented. To do this requires sophisticated instrumentation, and the workshop ended on a high note, with the promise of exactly that being delivered by HERMES on the AAT. A rather lively discussion as to exactly which individual chemical elements should be probed by this exciting new instrument inspired confidence in the community's ongoing zest for a better understanding of every single nook and cranny of our Galaxy.

As Tim Beers of Michigan State University, and Daniela Carollo of Macquarie University, in Sydney, concluded in their wrap-up of the workshop, published in *Nature Physics*, 'May the great barriers to Galactic Archaeology continue to fall, and the magnificent Great Barrier Reef, from which the conference drew its name, continue to live on'. Clichéd or not, I think we can all drink to that.

HOME, SWEET HOME

Perhaps it seems a bit odd that our home in the Universe—the Milky Way Galaxy—should be one of the last places to succumb to detailed probing by astronomers. Arguably, both our Solar System and the Universe at large are better known to science. The former is within our reach for physical examination with spacecraft, while we can observe the latter in directions away from the dusty disc of the Galaxy. But this exploration of our cosmic home is a difficult mission whose time has come—by courtesy of innovative astronomical instruments and sophisticated computer models. I feel privileged to have been involved in a small way, starting with observations of stars in the Galactic bulge in the late 1970s (when anything that wasn't cosmology was terribly unfashionable among astronomers) and now participating in the RAVE survey.

Like the hypothetical bacterium on its way to Pluto, thinking homely thoughts about planet Earth, those of us in the field of Galactic archaeology feel warmed by the knowledge that we are learning ever more about our true home in space—the Milky Way Galaxy. Even if we'll never be able to take a photograph of it from the outside, at least we'll know how it got to be the way it is.

And, finally, just in case you do happen to be a Pluto-bound bacterium, here's a word of advice. Watch out for the 10.9 kilograms of radioactive plutonium dioxide that you're sharing the spacecraft with. As I mentioned earlier, solar panels aren't much use in an environment that receives only one 1000th as much heat and light as Earth, so you'll need its warmth to keep going. Just don't get too close.

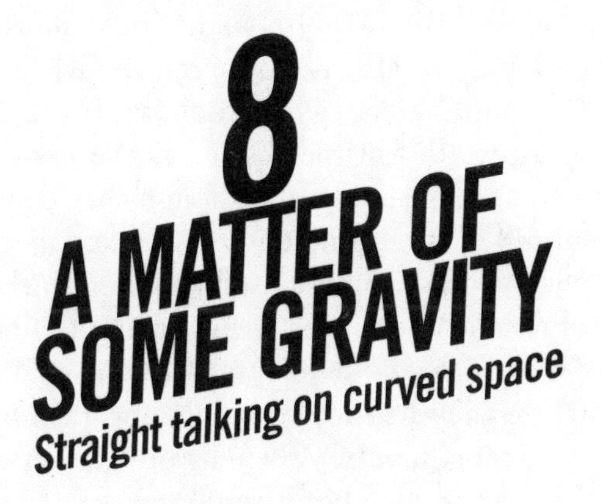

8 A MATTER OF SOME GRAVITY

Straight talking on curved space

There can't be too many book chapters that come with a government health warning. This one does—but I'll try to deliver it as gently as I can. If you suffer from arachnophobia, it might not be a bad idea to skip the first few paragraphs. In fact, don't even peep.

Australia is famous for its spiders, of course—although in reality you don't see that many more here than you do anywhere else. It's just that, here, they tend to be a bit bigger, are often meaner-looking and are sometimes vaguely poisonous. Very occasionally, they are lethal. But in Australia, as elsewhere, you're far more likely to die in a road accident than from a spider bite.

So, what prompted this excursion along the byways of Australian spider lore in a chapter supposedly exploring the byways of gravity? It comes from an event that occurred several years ago, when my decidedly arachnophobic twelve-year-old son awoke one summer morning to find a large and extremely handsome huntsman spider on the wall opposite his bed. These sleek-looking creatures have a smallish body sprouting what seem like dozens of long, hairy legs—which, on this one, measured a healthy 10 centimetres across their widest span. Huntsmen aren't venomous, although they can give you a painful nip if, for example, they've taken up residence inside one of your gumboots and you innocently attempt a spot of double occupancy with your foot. They spend most of their working lives spread-eagled against whatever surface they happen to be on, and feed by simply grabbing passing insects. When disturbed, they can move with astonishing speed on those long legs, but they do tend to ignore anything that's happening away from the surface on which they're sitting.

Given James' arachnophobia—and the fact that he was on the brink of his attitude years—I have to give him great credit for not leaping out of bed shouting 'Oh my God, there's a huge —ing spider in my room!' at his usual waking time of 6 am. Instead, he waited until a more reasonable hour before calmly alerting his morning-challenged father. Assisted by James' non-arachnophobic younger brother, Will, I then removed the spider by the time-honoured method of gently covering it with a large transparent kitchen bowl, sliding a sheet of thin card between it and the wall, and delivering it somewhere more distant than the bedroom—to wit, the farmer's field over the garden fence.

But the spider looked far from happy about this. In fact, it appeared to be shaken rigid—and that is when it occurred to me that we had just committed a horrendous crime against the natural order of spiderhood. Huntsmen normally inhabit the purely two-dimensional world of a surface—a comfortable bedroom wall, for example—and we had just propelled this one into a three-dimensional nightmare of coarse grass in Fred Swanson's back paddock. No wonder it was peeved. Suddenly having an extra dimension to deal with would be enough to give any huntsman spider a hefty dose of agoraphobia.

SPECIAL CIRCUMSTANCES

If it's any consolation to the spider, something similar happened to humankind rather more than a century ago. For over 200 years, since the time of Isaac Newton, we had assumed that we occupied a universe of three dimensions, and then along came a fellow called Albert Einstein who said no, there's actually another one. Einstein was not the first to suggest that time might somehow behave like a fourth dimension, but the idea is so deeply associated with his Special Theory of Relativity of 1905, that we tend to give him the credit. Special relativity is Einstein's theory of motion, which extends the ideas from Newton's *Principia* into a new realm—one in which objects are moving close to the speed of light. It was, in fact, Hermann Minkowski, a young professor at the University of Göttingen, in Germany, who finally gave us the full-blown four-dimensional Universe in elegant mathematical form, three years later. Far from being independent, said Minkowski, space and time were intertwined in a four-dimensional entity called 'space-time'. Sadly, within months of this coup, Minkowski died from a burst appendix, at the age of only 44.

A key difference between Newton's Universe and Einstein's is that Newton thought of time as an absolute—something that was meted out at the same rate everywhere and against which everything else was measured. There were no awkward questions about simultaneity, for example; if two events happened at the same time anywhere in the Universe, they were simultaneous. But Einstein asked what would happen when simultaneous events were seen by two observers, one of whom was moving with respect to the other. He found that, because of the finite speed of light, the notion of simultaneity gets rather vague. Things that happen together for one observer are separated in time for another.

If that seems hard to visualise, try swapping the concept of time for that of space—because another feature of Einstein's brave new world is that the four dimensions are subtly interchangeable. Let's imagine you're sitting on the upper deck of a Qantas super-jumbo en route from Sydney to London to join a Fred Watson study tour of spider habitats around historic European observatories. You've finished reading the latest issue of *Australasian Arachnid Fanciers Monthly*, and dinner is about to be served. Half an hour later, as you take your last sip of lukewarm coffee and digest your Qantas-issue Tim-Tam, you reflect on the fact that two events—the beginning and end of your meal—have occurred in exactly the same place, since you haven't moved from your seat. But to an observer on the ground, those events are separated by nearly 500 kilometres.

The occurrence of events in space and time is therefore relative rather than absolute—hence the name of Einstein's theory. And it's 'special' because it refers to the special case of things moving with respect to one another at a constant speed, as in an airliner at its cruising altitude. It has to be said that the discovery that we live in a four-dimensional Universe was nowhere near as traumatic for us as it would be for a huntsman spider suddenly finding itself transferred from a two- to a three-dimensional world. Our extra dimension is not populated with swooping magpies looking for a quick meal. But space-time did take some getting used to—especially when people realised what it could lead to.

Einstein's Special Theory of Relativity is based on two surprisingly simple postulates. The first is that the laws of physics are the same whether you're moving or not. You can reflect on this, too, as you finish your airline meal. Unless there's turbulence over Australia's top end, your coffee stays in its cup, even though you're moving at nearly 1000 kilometres per hour over the ground. The other postulate, while still simple, is a little harder to get your head around. It is that the speed of light in a vacuum is always the same, no matter how fast the source of the light is moving. This completely unexpected result was demonstrated in the 1880s in a famous piece of work called the Michelson-Morley experiment. Performed, not surprisingly, by physicists Albert Abraham Michelson and Edward Williams Morley, the experiment was successively refined by those gentlemen over the next 40 years, but with no change in the outcome. No matter how fast the light source is moving, the light it emits always travels through space at 299 792.458 kilometres per second (or 300 000 kilometres per second as near as makes no difference).

When Einstein wove his mathematical magic around these ideas, he arrived at some unusual results about time and space that take on real significance for objects moving at close to the speed of light. To a stationary observer, for example, an object whizzing by at high speed will appear slightly shortened along its direction of travel, a phenomenon called Lorentz-FitzGerald contraction (after—you've guessed it—physicists Hendrik Antoon Lorentz and George Francis FitzGerald). If you're thinking of using this trick to impress your friends with a new slimline you, however, forget it—it only works if you're travelling close to that 300 000 kilometres per second. And another, less welcome effect of special relativity will also kick in—your mass will increase as you approach the speed of light.

Perhaps more unexpected than the changes in geometry and mass is that time is also affected by your speed. To the stationary observer (who clearly has extraordinarily good eyesight), your watch will appear to slow down as you approach the speed of light. This phenomenon of time dilation immediately dispels the notion of a universal timescale. The fact is that all of us take our individual clocks with us, and the rate at which they tick depends on our state of motion. Of course, most of the human race is moving at roughly the same speed, certainly in comparison with the speed of light, so, for most practical purposes, we all experience time passing at the same rate. But today's most accurate clocks can easily detect the tiny effect of time dilation in spacecraft, and even in aircraft.

Dramatic and unexpected though these consequences of special relativity were, they were eclipsed by its most significant prediction—certainly as far as everyday life was concerned. Almost as an afterthought in his work on the Special Theory, Einstein arrived at perhaps the most famous equation of all time, linking the mass of an object with its value in energy. You know the one I mean. It is this equation that allows us to understand the prodigious energy output of stars and to contemplate the generation of atomic energy—as well as predicting the devastating power of nuclear weapons. These energetic processes involve reactions between the nuclei of atoms, and result in matter being converted directly into energy. But Einstein's equation takes the quantity of mass and multiplies it by the square of the speed of light—a truly enormous number—to get the equivalent energy. Hence their prodigious output.

In 1905, those particular discoveries were in the future. In fact, for Einstein, almost everything was in the future. Having had a rather undistinguished academic career, he was, at the age of 26, working as a patent clerk in the Swiss capital of Bern. He had acquired the job, in 1902, on the recommendation of a friend's father, having relinquished his German nationality for Swiss citizenship the previous year. And his job had just been upgraded from a temporary to a permanent position—a boost for a young family man seeking financial security. What was more unusual about this technical expert (third-class) was that in his spare time he tinkered around with theoretical physics. And he was simply brilliant at it.

Today, the city of Bern relishes its connection with this highly productive period of Einstein's life. Crowds flock to see his house at 49 Kramgasse—or perhaps it's to visit the coffee shop in the arcade below, though they can't fail to notice the life-sized portrait of the great man in one of the first-floor windows. But if you have any interest whatever in the life and work of this most remarkable scientist, you can do no better than visit the superb Einstein Museum, located in Bern's Historical Museum, just across the Aare River from the minster. There, you can experience every detail of Einstein's life—his strong Jewish roots, his education and early work, his famously philandering love life, his pacifism, and the progress of his career until his death, in 1955, at the age of 76. The exhibition is stunningly presented and includes hallowed artefacts of Einstein's life and work, many of which are moving in their intimacy.

It's not without reason that we tend to refer to 1905 as Einstein's *annus mirabilis*—his miraculous year. In fact, special relativity was just one aspect of what amounted to a single-handed revolution in physics. Here's the timeline:

On 17 March, Einstein submits a research paper on the photoelectric effect and the quantum nature of light, suggesting that light comes in the bullets of energy we now call photons (he won the 1921 Nobel Prize for this work). On 30 April, he completes a paper on the size of molecules (for which he receives his PhD from the University of Zürich). On 11 May, he submits a paper on the motion of small particles suspended in a liquid, leading eventually to the proof that atoms exist. On 30 June, he submits the paper on his new theory of motion, the Special Theory of Relativity. And on 27 September, he submits a short supplementary note on some consequences of special relativity—including *that* equation.

Given such a breathtaking performance, there is little wonder that the world celebrated the centenary of Einstein's *annus mirabilis* with an International Year of Physics. A worldwide program of events in both science and the arts included conferences, lectures, displays, exhibitions, concerts and plays. It was also the year in which the exhibition now in the Bern Einstein Museum was launched. As a result, 2005 was an outstandingly successful Einstein Year. Its one downside was that many people involved with the celebrations felt completely Einsteined-out by the end of it. And that is a pity, because, in reality, Einstein's *annus mirabilis* was just the start.

GRAVITY—OR ITS EQUIVALENT

By 1907, Einstein had a couple more publications under his belt and was beginning to win some recognition in scientific circles, although his application for a post at the University of Bern that year was turned down. He was still at the Patent Office. But it was while he was sitting there at his desk, no doubt musing about life, the Universe and everything, that Einstein had what he later described as '*der glücklichste Gedanke meines Lebens*' (the happiest thought of my life). Given that this man had some quite spectacular thoughts from time to time, his happiest one would have to be something pretty good.

Einstein had been thinking about how special relativity might modify Isaac Newton's famous theory of gravitation, which we met back in Chapter 5. Arguably the greatest intellectual feat of all time, Newton's theory had proved amazingly successful in explaining the motions of objects in the Solar System down to the finest detail. It had been instrumental in French astronomer Urbain Le Verrier's prediction of the existence of a planet beyond Uranus based on observed irregularities in Uranus' orbit—calculations that had led to the triumphant discovery of Neptune, in 1846. Newtonian gravity still works satisfactorily today in most of the questions associated with orbital motion and the navigation of spacecraft between planets. One day, it could even save the world from a threatening asteroid, if the offending object could be given a slight sideways tug by the gravitational attraction of a massive (20 tonnes or so) spacecraft.

By the end of the nineteenth century, Newton's gravity seemed able to explain everything, except one small issue. Back in 1859, the doughty Le Verrier had noticed a minute error in the predicted behaviour of Mercury's orbit. A tiny fraction of the steady sideways drift of the orbit could not be accounted for by Newtonian mechanics. This highly esoteric problem was enough to send poor Le Verrier off on an entirely understandable tangent, looking for an imaginary planet called Vulcan between the orbit of Mercury and the Sun. He didn't find it—mainly because it doesn't exist. Apart from that one minuscule aspect of Mercury's orbit, however, the Solar System was completely understandable in terms of Newton's theory of gravitation. But Einstein wondered how relativity might affect it, given that Newtonian gravity took no account of time—and he now knew that time and space were intimately linked.

Perhaps because he was bored to death at the Patent Office, Einstein then imagined himself falling from the roof of a tall building and realised that in this rather inconvenient circumstance he would feel no gravitational force. He would certainly feel something when he hit the ground, but that didn't matter in his thought experiment. Einstein reasoned that he would be in a state of free fall, and if his pipe fell out of his mouth or coins fell out of his pocket they would appear to float around him as if there were no gravity.

We inhabitants of the 21st century are, of course, used to seeing TV images of orbiting astronauts surrounded by the weightless detritus of their trade—floating pens, cameras, food capsules, scientific instruments, sick bags and so on. They, too, are in a state of free fall, but they never hit the ground because the forward motion of their spacecraft matches the rate at which the ground falls away beneath them—another possibility that had been spotted by good old Newton back in 1687. Incidentally, you can exactly replicate this weightlessness yourself, not by jumping off a tall building—unless you've truly had enough—but by jumping onto a trampoline. During the one second or so of each jump that your feet aren't touching the mat, you're as weightless as an astronaut, and you can prove it by watching coins or keys float from your open hands as you jump. It's a lot cheaper than going into space.

Commonplace though weightlessness is to us, it was an entirely novel idea for Einstein, and it quickly led him to the next step in his thinking: the realisation that the effects of a gravitational force and an applied acceleration are identical. In the case of your imaginary high jump from the top of a building, the downward acceleration you experience exactly negates the downward pull of gravity, so you become weightless. Looking at it another way, if you were sitting in a windowless compartment on a rocket deep in space and someone lit the fuse, you would not be able to tell whether the force you felt was due to the acceleration of the rocket or to gravity. Therefore—locally, at least—they must be the same thing.

This happiest thought of Einstein's life was a major breakthrough and is now called the Principle of Equivalence. In fact, it wasn't until 1912 that Einstein set out a formal statement of the principle, putting it in terms of the equivalence of the gravitational mass of an object (the way it responds to the pull of gravity) and its inertial mass (the way it responds to a force like the thrust of a rocket).

Einstein then began looking for the mathematical tools he needed to develop a new theory of gravity based upon the idea that it was equivalent to acceleration. In particular, he imagined a reference frame (a system of coordinates like the x, y and z we met in Chapter 7) being accelerated, with the observer inside it—thereby placing the observer within an accelerated reference frame. If you're not sure whether you know what an accelerated reference frame feels like, imagine yourself back in the Qantas super-jumbo. When it makes its take-off run with those four giant turbofans at full power, you, and everyone else on board, are in an accelerated reference frame. It's unmistakeable, and quite distinct from the constant-velocity reference frame you'll experience when drinking your coffee at cruising speed.

It is fairly well known that Einstein's deep insights into physics were not entirely matched by his abilities in mathematics. He was highly competent, of course, but relied on others to channel his thoughts in appropriate directions. Fortunately, his Special Theory of Relativity had brought his name to the attention of many leaders in the field, so he had a wide circle of friends and colleagues upon whose sharp minds he could call. One of those was a remarkable mathematician and astronomer by the name of Erwin Freundlich, a man whose role in the story of relativity is often underrated. Like Einstein, Freundlich was German born, but, unlike Einstein, he'd already had a spectacular academic career, culminating in receiving a doctorate from the University of Göttingen in 1910 when he was 25. He then worked at the Royal Observatory in Berlin, and it was there that he began collaborating with the great man. To start with, he obtained some accurate observations of Mercury to try to establish whether Einstein's new ideas had any bearing on the problem of its orbit. But then he unwittingly made an even bigger contribution.

Einstein's thinking on the equivalence principle and its relevance to gravity had led him to the view that much of it boiled down to a problem of geometry—because the laws of relativity that applied in ordinary, or Euclidean, space didn't work in an accelerated reference frame. He needed a new kind of geometry in which the shape of space itself was modified by forces—called 'fields'—acting through it. Musing aloud on this to Freundlich one day, Einstein was astonished to be told that such a complex model of space had been known to mathematicians for over half a century. It had been developed in the 1850s by another gifted German mathematician, called Georg Friedrich Bernhard Riemann, and was known as Riemannian geometry. But it involved some *very* difficult mathematics. According to Walter Ledermann, a later colleague of Freundlich's, Einstein was so amazed by this news that he accused Freundlich of lying. But it was soon proved to be true. Poor old Einstein then had no alternative but to grasp the painful nettle of Riemannian geometry and begin slogging his way through the algebra.

RELATIVITY COMES OF AGE

What followed was the General Theory of Relativity, so called because it wasn't limited to the special case of objects moving at a constant velocity. It amounted to a startling new theory of gravity that made the wildly improbable assertion that space-time itself can bend, warped by the presence of matter—which, in turn, responds to the distorted geometry of space-time by moving within it. In practical terms, what that means to you and me is that the downward pull we feel at the surface of the Earth is not, in fact, due to a force existing between the Earth and ourselves, as Newton had proposed. Rather, the shape of the space around us changes very slightly between our feet and our head due to the presence of the Earth. We feel that change in shape as gravity. Gravity is therefore a property of the Universe itself rather than a property of objects within the Universe.

That bland description gives little hint of the complex mathematics needed to describe the theory. Riemannian geometry requires a tool called 'tensor calculus', which, to mathematically challenged individuals like me, spells the utmost in doom and gloom. However, the final step in the argument was rather nicely put into words by the late British astronomer and mathematician Sir Fred Hoyle in a popular article in the 1980s:

> Einstein's remarkable idea was to regard the difference between Riemannian spacetime and [Hermann] Minkowski's spacetime as the true meaning of the phenomenon of gravitation. To this end he modified Newton's equations of motion so as to form a comprehensive scheme for calculating not just the motions of particles in a prescribed spacetime like that of Minkowski, but a determination of what the more complex Riemannian spacetime had to be.

Not surprisingly, by the time Einstein submitted his work for publication, on 25 November 1915, his mathematical endeavours had taken him through several versions of the theory. Most of these he had published, which meant that other scientists were fully aware of what he was up to and where his thinking might lead. In particular, another able German mathematician—a man seventeen years Einstein's senior—was working hard on the same theory. This was David Hilbert, who, some have claimed, was robbed of glory by Einstein, because he had actually submitted the correct version of the general relativity equations for publication five days before Einstein did. It seems more likely, though, that Hilbert and Einstein were bouncing ideas off each other, since they were writing cordially to one another throughout that heady November.

A week before submitting his work, Einstein had a tremendous confidence boost. He realised that the final version of his theory exactly accounted for the observed anomaly in Mercury's orbit, for so long a thorn in the side of astronomy. 'For a few days, I was beside myself with joyous excitement,' he wrote later. And who could blame him?

By the time Einstein completed his General Theory of Relativity, he was once again living in Germany and working at the University of Berlin. He was now close to Freundlich, with whom he had long discussions on how to test the theory. This idea of space-time being warped by solid objects was so outlandish that it risked being laughed out of court by most of the astronomers of the day. Newton's ideas had stood up extremely well for two centuries and had almost acquired the air of religious dogma. To challenge them was dangerous. Not to life and limb, of course—persecution had come a long way since the seventeenth century—but to one's career. What was needed was a critical test of the theory, and Freundlich was Einstein's right-hand man in planning it. One of the predictions of general relativity is that a massive object such as the Sun will distort space enough to bend rays of light passing close to it, rather in the fashion of a glass lens. So, if the Sun were a dark object rather than a bright one, you would be able to see the apparent positions of stars close to the edge of the Sun being deflected very slightly away from it as the Sun slowly moved through the sky in front of them. In practice, there's only one way to make the Sun dark, and that is to hide its light by having the Moon pass in front of it. Since that is exactly what happens in a total eclipse of the Sun, the supporters of Einstein's general relativity suddenly became avid eclipse-chasers. For them, astronomy tourism took on a quite different meaning from the leisurely study tours featured in this book.

And the ever-faithful Freundlich was first off the mark. While the General Theory was still in an incomplete form, he secured funding to mount an expedition to Feodosiya, in the Crimea, to observe a total eclipse late in 1914, with the aim of testing the theory. But his timing was awful. Before the eclipse occurred, hostilities were declared, and the First World War lumbered forth on its dreadful course. The eclipse expedition was abandoned, but it was too late. As the holder of a German passport, Freundlich was interned in the Crimea as an enemy alien.

PEACEMAKER

Meanwhile, on the other side of the world, the British Association for the Advancement of Science had been holding the various sessions of its annual meeting across five cities in Australia. With nothing remotely like today's wide-bodied jets to ferry the participants around, this was a remarkably ambitious step for a scientific organisation. The meeting had gone well but had been marred by the declaration of war, received while the sessions were in progress. Not surprisingly, this had led to the early embarkation of many of the participants on steamers back to the Mother Country. Before they left, however, they had earnestly resolved to maintain business as usual, asserting that 'science is above all politics'.

Within weeks, though, that assertion counted for nothing. Researchers on both sides of the Anglo–German divide had followed the general public's slide into unashamed jingoism. When no fewer than 93 prominent German scientists signed a *Manifesto to the Civilised World* defending Germany's 'struggle for existence', their British counterparts hurled accusations of complicity in barbaric atrocities and set about excluding the Germans from all normal channels of scientific communication.

Nowhere was this unseemly crusade mounted more intensively than in the world of astronomy, where impassioned voices on both sides of the conflict asked whether there could ever again be normal relations between the two communities. The professional astronomy journals of the time, still available today in the libraries of many long-established observatories and university astronomy departments, reported on British astronomy meetings at which strident accusations of war crimes were made against German scientists. Prominent among the accusers was Herbert Hall Turner, uncompromising anti-German professor of astronomy at Oxford University, who we met in a very different light in Chapter 2.

As the war dragged on, attitudes became ever more hardened, and long-established academic ties were discarded like spent shell cases. At the Armistice of 11 November 1918, it seemed as if decades must pass before scientific relations could be normalised. Yet within a year the British press was proclaiming a German-born scientist as the champion of the age—someone, moreover, who had overthrown the hallowed ideas of the local hero Sir Isaac Newton. Soon afterwards, scientific cooperation between the two former enemies began to be re-established.

The German-born scientist was, of course, Albert Einstein. When his General Theory of Relativity had been published, in the dark days of 1916, the shut-down in international communication meant that it was accessible only to German scientists. However, one reader in neutral Holland was convinced that British scientists, too, needed to know about it. This man was an astronomer, Willem de Sitter, who still had scientific contacts in the United Kingdom. In the event, though he was uncertain who might read it, de Sitter sent his own interpretation of general relativity to the United Kingdom. It found its way to perhaps the only British scientist who was both capable of grasping its significance and open-minded enough to read work that had originated in wartime Germany. This was Arthur Stanley Eddington, director of the Cambridge Observatories and an astrophysicist of outstanding ability. More importantly, he was, like his German-born counterpart, an ardent pacifist. In fact, that pacifism came from Eddington's strong Quaker beliefs, which were, for him, as much a framework for life as Einstein's Jewish background.

Throughout the latter half of the war, Eddington studied the new theory in great detail and recognised the genius behind it. Like Freundlich, he became a champion of relativity, seeking ways in which the theory could be proved by observation. There is a delightful story of Eddington being congratulated by a colleague for being one of only three people in the world who understood relativity. When Eddington paused, and his colleague commented that there was really no need to be so modest about it, the Cambridge astronomer replied that, no, he was just trying to think who the third person might be.

As is well known, Eddington was instrumental in setting up the eclipse expeditions that became the litmus test of general relativity. On 25 May 1919, a solar eclipse in fortuitous circumstances was predicted. It would take place against the backdrop of the Hyades star cluster—a part of the sky rich in bright stars that were perfect for the measurement of their light's deflection by the eclipsed Sun. The path of the eclipse was in the southern hemisphere, but two British expeditions were organised to well-separated sites on the path, one on the island of Príncipe, off the west African coast, and the other in Sobral, in Brazil. Eddington led the team in Príncipe.

Einstein's prediction was that light from the stars near the edge of the Sun's disc would be displaced by the tiny angle of 1.75 arcseconds. An arcsecond is one 3600th of a degree, but I always find it more instructive to imagine someone holding up a $1 coin at a distance of 5 kilometres. The coin, from the observer's point of view, is 1 arcsecond in diameter—and completely invisible to the observer's unaided eye (as is the person holding it). The unlikely sounding feat of measuring 1.75 of those minuscule angles was already standard practice in the photographic astronomy of the day, so promising results were obtained from both eclipse sites. There was then a lengthy process of measurement and calculation, which took place back in the United Kingdom.

At a combined meeting of the Royal Astronomical Society and the Royal Society in London on 6 November 1919, Eddington finally announced the results. Yes, the prediction of a 1.75-arcsecond displacement was correct, and Einstein was nothing short of a genius. Eddington took pains to ensure that this triumph of international collaboration was properly reported in the world's media. He need not have worried. *The Times* of 7 November 1919 blared: 'Revolution in Science—New Theory of the Universe—Newtonian Ideas Overthrown.' None of which was any exaggeration.

The revolutionary new theory of gravity, in which space and time are warped by the presence of matter, had been proved at a solar eclipse observed in far-off lands. Moreover, a comparatively unknown Englishman, motivated by a deep-seated belief in peaceful international collaboration, had provided the dramatic confirmation of the German-born Einstein's theory. As a result, Einstein himself immediately shot to world fame, but the consequences went much deeper. It can reasonably be argued that, between them, Einstein and Eddington forged the peace that quickly took hold in the scientific world. The developing entente helped to put relativity in the public spotlight, where it remains firmly to this day as the most important theoretical foundation of our understanding of the Universe.

Despite the starring roles of Einstein and Eddington, there were two other important players in this drama. The first was a brilliant young mathematician by the name of Karl Schwarzschild. Within weeks of the publication of Einstein's final paper on general relativity, Schwarzschild had solved the equations that represented the gravitational effect of a massive compact spherical object. At the time, this was purely of theoretical value, but when astronomers began taking an interest in such things, in the 1960s, Schwarzschild's mathematical solutions assumed vital importance. Today, we call these objects black holes. Sadly, no triumphant accolades awaited Schwarzschild; he died as a result of illness contracted on the Eastern Front, in 1916.

And what of our old friend Erwin Freundlich? Fortunately, his internment in 1914 was short lived, and he was soon able to return to Berlin. There, he continued his work on observational methods for demonstrating the validity of relativity, working closely with Einstein. In their spare time, the two played music together, Einstein on violin and Freundlich on cello. Freundlich seems to have shown no resentment whatever when the British eclipse expeditions provided the first proof of general relativity, and he went on to mount several more eclipse expeditions of his own, to improve the accuracy of the determinations. He accompanied Einstein on a visit to the United Kingdom in the early 1920s and was feted alongside the great man.

In 1921, Freundlich was appointed to the newly created Einstein Institute at the Astrophysical Observatory in Potsdam. There, he was able to explore one of the other predictions of general relativity, that to an outside observer clocks in a gravitational field appear to run more slowly. This is nothing to do with faulty clocks; time itself is slowed because of the effect of gravity. It is called 'gravitational time dilation', and if it sounds familiar, well spotted—it's analogous to the speed-related time dilation mentioned a few pages ago. Today's super-accurate atomic clocks are able to detect the difference in time caused by the change in the Earth's gravity between the top and bottom floors of a skyscraper. But in Freundlich's day, that was far in the future. The only possible way of investigating it was via a phenomenon called 'gravitational redshift', in which the barcode information in the light leaving a massive object is shifted slightly towards the red end of the spectrum. This is a direct consequence of gravitational time dilation, and if you could explore the spectrum of the

Sun in fine enough detail, you would have a chance of measuring it.

Thus, Freundlich was instrumental in creating the Einstein Tower—the Expressionist tower telescope in Potsdam that we visited at the beginning of this book. The idea was that a system of mirrors at the top of the tower would guide a magnified image of the Sun to the bottom, where its light could be dispersed into a detailed spectrum. Unfortunately, this monument to relativity never did what it was built to do—the detection of gravitational redshift is an extremely difficult observation and was not finally achieved until 1976.

The Einstein Tower caused a stir of a quite different kind when it was visited by members of the Stargazer II tour in 2010. When the group was told by our guide, Matthias Steinmetz, that Einstein himself had been a regular participant in discussions around the conference table on the tower's ground floor, tour members wanted to know in which chair the great man had sat. Smiling, Steinmetz told us it could have been any of them—whereupon two-dozen erudite tour participants became a bunch of school kids, taking it in turns to sit in *all* the chairs, just to make sure. And, yes, your humble author was one of them.

A RELATIVELY SAFE HAVEN

The dramatic political changes that forced Albert Einstein to forsake Germany for the United States, in 1933, also affected Erwin Freundlich. Though not a practising Jew himself, he and his wife were of Jewish descent, and so they, too, left when the Nazis came to power. Moving first to Istanbul and then to Prague, Freundlich continued his work in astronomy. But with Hitler's advance on Czechoslovakia, the Freundlichs were forced to move again, finally coming to rest in 1939, in the relative safety of Scotland, at the University of St Andrews.

Famous today as the *alma mater* of the Duke and Duchess of Cambridge—alias Wills and Kate—St Andrews is the oldest university in Scotland and among the oldest in Europe, having received its papal bull of foundation in 1412. It has a long tradition of mathematical astronomy going back to James Gregory, in the 1670s, as we saw in Chapter 5. But in 1939, when Freundlich arrived, there was no astronomy department. Arthur Eddington recommended to the university authorities that such a department was needed—and that Erwin Freundlich was exactly the right person to create it. Thus, Freundlich embarked on a phase of his career in which he was exceedingly happy. Just like William Herschel, 182 years before, he was quick to integrate into the culture of his adopted country and took on a new name to prove it. Since his mother's maiden name had been Ellen Elizabeth Finlayson—and there are few names more Scottish than that—he duly became Erwin Finlay Freundlich and was known by that name for the twenty years he lived in Scotland.

Having successfully founded the astronomy department, Finlay Freundlich went on to build a new observatory in open land surrounded by the university's playing fields, equipping it with the most modern observing facilities. Indeed, his plans in that regard were hardly modest, including a proposal to build the most powerful optical telescope in the United Kingdom. A half-scale model was built first, a 38-centimetre-aperture instrument named the Scott Lang Telescope, after a benefactor to the university. In fact, the full-scale version, the 76-centimetre James Gregory Telescope, was not completed until after Finlay Freundlich left St Andrews in retirement, in 1959. For some years, however, it was indeed the most powerful telescope in the United Kingdom.

Perhaps even more important than equipping the fledgling observatory was Finlay Freundlich's vision of attracting capable astronomers to its staff. This was considerably enhanced when he was appointed to the newly founded position of Napier professor of astronomy, in 1951. But among his earlier recruits was a young Polish refugee by the name of Tadeusz Boleslaw Slebarski, one of Finlay Freundlich's protégés in mathematical astronomy, although not himself a specialist in relativity. Like Finlay Freundlich, Slebarski was a well-respected man, gentle in manner and generous in spirit. He was known throughout the university simply as Mr Slebarski, and his lectures were models of clarity because, having a less-than-perfect command of English, he wrote absolutely everything on the blackboard. Someone once unkindly remarked that he had forgotten all his Polish and had never learned anything to replace it—but that vastly underrates his capacity as a teacher of science.

Slebarski has a special place in my own history, because a quarter of a century after his appointment—and a decade after Finlay Freundlich had retired to Wiesbaden—he became my research supervisor when I embarked on a masters degree at St Andrews. We worked not on relativity but on asteroid orbits, and I used the Scott Lang Telescope to make my observations. With his customary good nature, Slebarski tolerated the fact that I was far from a model student. I still remember his delight when, years later, I finally submitted my thesis and graduated with the degree.

Neither Finlay Freundlich nor Slebarski is alive today, but their memories live on. For me, they form a direct academic link with Einstein—something of which I would be very proud if I felt I had any right to be. However, I have to come clean and admit that it's only in recent years that I've properly understood relativity. In my student days, I'm afraid I simply didn't get it.

THE HUNT FOR NEW PHYSICS

Almost a century after its formulation, general relativity is still the best theory of gravity we have. It has survived all the tests that have been thrown at it and has successfully predicted the existence of black holes, the expanding Universe, gravitational lenses, gravitational waves and many other phenomena dear to the hearts of astronomers. It underpins cosmology—the science of the history and evolution of the Universe as a whole. Many non-scientists are familiar with the idea that space-time is curved, and Einstein is regarded as the greatest genius of the twentieth century—not to mention a few other centuries besides.

The one thing that general relativity can't cope with is gravity on the smallest scale, since the theory is based on space-time as a continuous medium rather than a succession of distinct steps, or quanta—and we know that this is how the submicroscopic world works. The search for a theory of quantum gravity akin to our understanding of the other fundamental forces of nature on submicroscopic scales is well underway. These forces include electromagnetism and two nuclear forces, and we will meet them again in Chapter 10. As yet, however, there is no clear leader among the many candidates for quantum gravity. In any case, these theories all rely on new physics—exotic underlying realities that are not predicted by relativity, in which physical processes as yet unknown are at work.

In fact, relativity itself offers a means of testing for the possibility of new physics. I mentioned a few pages back that the heart of the equivalence principle—and the heart of general relativity itself—is that the gravitational mass and the inertial mass of an object are the same thing. One of the consequences of the equivalence principle is that objects will fall under gravity with the same acceleration, no matter what their internal structure and composition. This harks back to Galileo's famous experiment at the Leaning Tower of Pisa, in the 1590s, and most of us have been familiar with the acceleration due to gravity at the Earth's surface (9.8 metres per second per second) since our schooldays. Sometimes from first-hand experience. But the bottom line is that some of the more exotic theories of gravitation—such as those attempting to unify the fundamental forces of nature—would, if correct, lead to a violation of the equivalence principle. Prominent among these unified theories is string theory, which postulates that forces and particles are manifestations of vibrating strings of energy. Hunting for subtle violations is therefore a great way to discover whether new physics is, in fact, at work.

How can one do this? The laboratory method involves comparing the responses of two differing test bodies to the same gravitational pull. Lumps of copper and aluminium sitting in the Earth's gravitational field, for example. The technique owes its origin to a gifted Hungarian nobleman, Loránd Eötvös, who performed such experiments with a sophisticated torsion (swinging) balance in the 1890s and early 1900s. His work demonstrated no difference in the acceleration experienced by differing test masses to within one part in 100 million, and it was this result that led to Einstein's adoption of the equivalence principle. The rather striking

Hungarian name Eötvös, by the way, is pronounced a bit like 'oat-fosh', which explains the whimsical name of the research collaboration that has taken this work forward into the 21st century. Its members are at the University of Washington, in Seattle, and are called the Eöt-Wash group. Boom boom.

Using extraordinarily sensitive balances, Eöt-Wash scientists have confirmed the equivalence principle to better than one part in a trillion using the gravitational attraction not just of the Earth but also of the Sun and our own Galaxy. In the late 1980s, they famously put paid to a flurry of excitement over the possible existence of a so-called 'fifth fundamental force', in addition to the four already known to science. Their experiments (and eventually those of other groups, too) showed no evidence of violation—and hence no fifth force. But even the spectacular accuracy achieved by Eöt-Wash is now being surpassed by more refined experiments, using light bounced off reflectors left on the Moon by *Apollo* astronauts, for example, and test masses in Earth orbit. The most ambitious are expected to provide answers with a head-spinning accuracy of one part in a billion billion.

Such fastidious explorations of the validity of general relativity via the robustness of the equivalence principle are comparable with the gigantic experiments that are now under way at facilities like the Large Hadron Collider, the famous European particle accelerator that straddles the French–Swiss border near Geneva. They, too, are looking for anomalies in our current model of the Universe. It has to be said that the Collider is the poster child in the search for unknown phenomena. But wouldn't it be astonishing if a whole era of new physics came from nothing more than the detection of a minuscule deviation from the equivalence principle, measured, perhaps, with an ultra-sensitive balance?

Just to put that into perspective, the probability of finding a deviation so impossibly small is about the same as discovering a solitary huntsman spider in a farmer's field. A farmer's field the size of Asia, that is.

9 GREENING THE UNIVERSE
Towards sustainable space science

Did you know that 2008 was the International Year of the Potato? I mention this not because I'm fond of potatoes—which I am—but because of what came next. The following year was the International Year of Astronomy, designated thus to celebrate the 400th anniversary of Galileo's first telescopic observations. We in the astronomy business were very keen to outdo the potato people in bringing our special subject to the widest possible audience. Whether we succeeded or not is for others to judge, but most of the world's professional astronomers, together with their amateur colleagues, did their bit in publicising their work throughout 2009.

My part in this was masterminded by the redoubtable Marnie, an expert in event management as well as a skilled tour coordinator. It included a range of enterprises, some of which were perhaps better advised than others, but all of which miraculously played to sell-out audiences. For example, there was an astronomy-themed magic show called *Time Warp*, featuring the acclaimed illusionist Matt Hollywood. (I wonder whether that's his real name?) He did all the clever stuff, while I tried to explain the science of time travel. Then there was an astronomy-themed jazz concert, *Hot Stars, Cool Jazz*, featuring the equally acclaimed trumpeter James Morrison, who wowed the crowds while I fooled around with inflatable planets. Naturally, the show included such favourites as 'Fly Me to the Moon', 'Stella by Starlight' and, well, 'Saturn Doll'. But in a different league—and the clear leader in job satisfaction—was an astronomy-themed expedition cruise around Australia's Kimberley coast. Ah, now you're talking. My colleague David Malin and I spent a couple of idyllic weeks bringing astronomy to 100 or so passengers aboard a small cruise ship aptly named *Orion* as she meandered her way from Darwin to Broome. Our astronomy lectures and late-night stargazing went down well enough, but it has to be admitted that what captured the hearts of all on board was the Kimberley coast itself. We encountered a completely unspoiled part of the planet—many of us for the first time.

I thought I was pretty familiar with wilderness areas, since the Warrumbungle Mountains are hardly the epitome of urban development. But even there you can always find signs of modern habitation, with power lines, radio towers and, yes, observatory domes dotted through the landscape. But in large tracts of the Kimberley there is not the merest hint of human presence. Not a pathway, not a telegraph pole, not a phone tower—nothing. Only the quickly vanishing wakes of the *Orion*'s Zodiac inflatables betrayed our exploration of this pristine world, as we marvelled at its geology, flora and wildlife.

The cruise provided me with food for thought. As I said somewhere near the beginning of this book, astronomers are inveterate travellers, and that doesn't come without a carbon footprint. And, as someone who advocates astronomy tourism, I have a carbon footprint that is vicariously bigger than most, because of the travels of our tour groups. Fortunately, all those folk are pretty environmentally aware and endeavour to limit their carbon emissions in other ways. And, of course, they represent a minute fraction of the travelling public. It begs the wider question, however, of 'Just how sustainable are the activities of astronomers—and, more especially, of space scientists?' When you look closely at that, some surprising facts emerge.

FIGHTIN' WORDS

For example, you may be surprised to learn that astronomers can get sucked into biopolitics. It doesn't happen often, but the 1990s saw the start of a bitter feud that still rumbles around today, at least among some of the United States' conspiracy theorists. At its heart was a diminutive hero, the 200-gram Mount Graham red squirrel, which is unique to the tree-covered Pinaleño Mountains in southern Arizona. The highest peak in this small, isolated range is Mount Graham, hence the squirrel's common name. It was thought to be extinct in the 1950s, but the plucky little chap reappeared in small numbers on the 3200-metre mountain during the 1970s. In 1987, with a population of about 300, it was formally listed as endangered. Serious though that might be, it was a lot better than being listed as extinct.

Just a year later, the US Congress authorised the construction of a major international observatory on Mount Graham. The University of Arizona, based in Tucson and home to one of the world's foremost astronomy research groups, had outlined an ambitious program of telescope building that would culminate in the Large Binocular Telescope, one of the world's biggest optical telescopes and the prototype for an even larger project in the southern hemisphere. It was not long before conservationists and astronomers found themselves at loggerheads over the future of the mountain. This was a situation of considerable discomfort to the world's astronomers, who are stung by any accusation of environmental vandalism. Observational astronomy's endeavours are underpinned by a deep awareness of environmental issues. Its scientists cherish wilderness areas as vantage points for exploring the infinitely greater wilderness beyond. Moreover, they are more inclined than most to regard the Earth as an extraordinarily special place in the Cosmos. They recognise that our planet's delicate balance of geophysical and atmospheric processes fosters an environment that is as fragile as it is unique.

Had the dispute centred only on the well-being of the squirrel, it would probably have been resolved relatively quickly. The Arizona astronomers were anxious that the telescope buildings should have the least possible impact on the squirrels' habitat, and every care was taken in the design of the observatory and related infrastructure such as access roads. Unfortunately, though, the environmental debate soon became entangled with another, equally sensitive issue, to do with the traditional ownership of the land. Mount Graham, like other peaks in the region, is a sacred site for the Apache people. While the mountain itself is hardly pristine, with holiday homes, visitor areas and educational camps occupying its flanks, the summit has remained largely unaffected. It was claimed that even minimal development of the mountain top for astronomy would desecrate it, violating the rights of the San Carlos Apaches to practise their religion and traditions.

Once again, it seems likely that compromise could have been reached quickly, had it not been for the attentions of an inflammatory press inciting a population that was, perhaps, genuinely fed up with having its access to local beauty spots restricted by scientific activities. 'Star Whores—Astronomers vs. Apaches on Mount Graham' raged one headline. A commentator in the *Phoenix New Times* took a more sanguine view, asserting that Arizona is the 'Home of the OK Quarrel', where feuding is just a way of life. 'Them's fightin' words,' he said, quoting with evident approval the 'star whores' comment and an astronomer's rejoinder that 'people and squirrels live together fine'.

Eventually, amid unsurprising accusations of coercion and foul play on the part of the University of Arizona, the Mount Graham International Observatory was given the green light by the US Congress, and construction went ahead. An acceptable compromise was reached with the Apache Tribal Council, and the university commissioned an independent census of the squirrel population. Surprisingly, it revealed a notable spike in squirrel numbers during the early construction phase of the Large Binocular Telescope, from 1998 to 2001. Since then, however, numbers have reverted to similar levels to those encountered before the bulldozers moved in and have stabilised at around 250 individuals. The current thinking among environmental biologists is that rainfall, cone crop, insect numbers and the occurrence of major forest fires (such as a particularly disastrous one in 2004) are more significant than human activity in the well-being of the squirrels. The Large Binocular Telescope, at least partially absolved, was eventually dedicated in 2004 and was fully operational by 2008.

SPACE INVADERS

The case of the Mount Graham squirrels is a rare example of astronomy openly falling foul of conservationists. But it does hint at a willingness on the part of astronomers to suppress environmental considerations in order to pursue their scientific goals. For their cousins in the space industry—and, in turn, the governments that want a space industry—that willingness seems to know no limits. Here, there are real instances of environmental vandalism.

The plight of farmers in the Altai Republic is a case in point. These unfortunate people live downrange of Kazakhstan's Baikonur Cosmodrome, from where Russian spacecraft are launched. They are variously bombarded with spent first-stage rockets, debris from failed launches and unburned toxic fuel. Why is the launch site in Baikonur? Space launch facilities are usually constructed as near as possible to the equator and with as much empty space as possible to the east. This is to take advantage of a remarkable free gift that our planet offers spacefarers.

In order to attain orbit, a spacecraft not only needs to be above the atmosphere at a height of at least 100 kilometres but also requires a spectacularly high velocity parallel to the ground to the ground—around 7.8 kilometres per second. At this speed, a spacecraft's fall back to Earth under gravity is negated, since the curved surface of the planet falls away at the same rate; hence, the craft stays in orbit. As I noted a few pages ago, this remarkable principle was first spotted by Isaac Newton, who envisaged firing cannon-balls into orbit. Today, the necessary ingredients of height and forward speed are provided by the launch vehicle, and the classic image of a rocket launch, with its initial vertical trajectory gradually leaning over towards the horizontal, is the obvious manifestation of that. It's not rocket science—despite what you may think.

But what about the free gift? By virtue of our planet's rotation, every point on its surface is being carried eastwards at a speed that varies with latitude. At the equator, for example, the surface is moving at 1674 kilometres per hour, or almost half a kilometre per second. If a spacecraft is launched from the equator in an easterly direction, that velocity adds to the horizontal speed provided by the rocket, yielding a 6 per cent discount on the velocity budget of the launch. This translates into a considerable saving in fuel and a reduction in the lift-off weight of the rocket. With the pump-price of rocket fuel what it is today, any saving is a good thing. Considerations of safety also suggest that your launch facility should have unpopulated country to the east, so that aborted launches and spent booster rockets—both of which involve debris falling back to Earth—will do minimal damage. At Cape Canaveral, that role is filled tolerably well by the Atlantic Ocean. Geography is not always helpful, however, and the old Soviet Union had to make do with less than ideal circumstances. In fact, the choice of Baikonur for the cosmodrome was almost accidental. In the 1950s, when the Cold War was in its infancy, the isolated town of Tyuratam in Kazakhstan was selected as the principal test site for intercontinental ballistic missiles, because it was surrounded by huge tracts of boggy, unpopulated territory in which radio guidance towers could be built.

Later renamed Baikonur, this became the Soviet Union's (and subsequently Russia's) major space launch facility. Even though its latitude is 46 degrees north, Baikonur's free gift still amounts to about 1160 kilometres per hour, so the velocity discount is 4 per cent.

Another, far less wholesome, legacy of the Cold War era comes in the shape of the Proton rocket, which quickly became the Soviet Union's workhorse medium-lift launch vehicle and is still central to the Russian space program. Proton is fuelled with nitrogen tetroxide and unsymmetrical dimethyl hydrazine, highly dangerous fluids that are both toxic and corrosive.

In the matter of unpopulated territory to the east, Baikonur is now sadly lacking. A few hundred kilometres downrange of the space centre is the Altai region of Siberia, whose farmers effectively live on the front line. Imagine a rural landscape in which, several times a month, blazing pieces of space debris rain down, spewing unburned toxic fuel. Through a classic piece of Soviet-era bureaucracy, the unfortunate occupants of an officially designated danger zone are given 24 hours' notice of possible impacts, but should damage or injury result they are not entitled to any compensation. That is reserved for folk living outside the fall-out zone.

The result of this has been increasing protest within the affected region. Locals complain not only of animal deaths through contact with toxic propellants but also of serious illness among the human population. Lamenting the passing of the good old days, they say that the efforts of authorities to clean up the landscape by removing contaminated debris have declined since the collapse of the Soviet Union. There is a flipside, however, in the burgeoning trade in scrap metal from rocket bodies, fuel tanks and external boosters. Altai is probably the only place in the world where farmers work the land with traditional implements fashioned from space-grade titanium alloy.

The space age has brought other environmental hazards that are probably better known than the plight of the Altai farmers. Twenty or so spacecraft have been launched with plutonium-based thermoelectric generators on board, most recently NASA's *Curiosity* rover on its mission to Mars. With failed launches in the past, nuclear electricity generators on board spacecraft are always going to be an extremely sensitive issue.

More familiar still is the accumulation of orbiting debris around the Earth. It ranges from flecks of paint (which famously chipped the space shuttle's windows due to the high relative velocity) to complete upper-stage rockets that remain in space after their payloads have been nudged into their operational orbits. There are around 800 active satellites in orbit, but a huge number of defunct components remain in space. US authorities (notably the North American Aerospace Defense Command) track more than 22 000 pieces of debris with dimensions above 10 centimetres, but uncatalogued pieces smaller than this are numbered in the millions.

There's no doubt that these objects pose a significant threat to the utilisation of space. Spacecraft of every kind, including those carrying humans, are under threat from them. While, in many cases, their orbits will gradually reduce in height (resulting in atmospheric re-entry and harmless burn-up), the timescale for this can be tens, hundreds or even thousands of years. Bingles involving spacecraft have already occurred. In February 2009, for example, a US communications satellite, *Iridium 33*, and a deactivated Russian satellite, *Cosmos 2251*, collided and disintegrated over the Earth's North Pole, producing large plumes of debris following the original orbits of the two spacecraft.

There are several cases in which particularly large pieces of debris have failed to burn up on re-entry and have found their way back to Earth. Fortunately, this has never resulted in a fatality, but the possibility is always there. A recent example was NASA's 6-tonne *Upper Atmosphere Research Satellite*, which reached the end of its useful life in 2005, but continued to orbit the Earth at a steadily decreasing height until it made an uncontrolled re-entry in September 2011. The surviving fragments landed in the Pacific Ocean.

Who is responsible for this pollution? The answer falls firmly within the province of space law, which is embodied in the United Nations-ratified Outer Space Treaty of 1967 and its four additional conventions of 1968–79. The problem is that these treaties were formulated in an era very different from our own, when the principal users of space were a handful of superpowers. Today, the exploitation of space has not only extended to the commercial sector but is also tempting the tourist industry, and the present laws are inadequate to deal with the possibilities that might arise. While the Outer Space Treaty is clear that any object in space remains the property of its original owner, the mechanism for claims against any damage it may cause is far from straightforward. There is a growing body of opinion that the treaty no longer matches reality, and it's high time it was reviewed.

It is only within the last few years that any serious thought has been given to the means by which space pollution might be reduced. Unlikely as it might sound, a promising suggestion is to equip all new space vehicles with a balloon approximately 40 metres in diameter, which can be inflated with helium once the craft has finished its useful life. The balloon would greatly increase the braking effect of the thin air in the upper atmosphere, bringing the satellite down to Earth in months rather than decades or centuries.

This doesn't solve the existing problem, of course. While a recent initiative from the Russian Space Agency might eventually result in a clean-up of defunct communications satellites at a height of 36 000 kilometres, there is no obvious solution to the problem of low-orbit space junk. The Swiss have proposed a suite of 'janitor' satellites to bring down large items, but political sensitivities make them unlikely to be deployed on any significant scale: if a space agency had the wherewithal to remove defunct satellites from orbit, what would prevent the same technology being used in a hostile act against operational spacecraft? The best we can hope for in cleaning up the Earth's environment seems to be for time to take its toll by the natural decay of spacecraft orbits, which eventually leads to them burning up in the atmosphere.

The decision to have space programs was made independently of whether scientists would use them; astronomers and space scientists merely took advantage of the opportunity. So, whether the resulting mess is in space or in places like the Altai Republic, who is to blame for it? Astronomers, space scientists, governments? Or the electorate? It's a tricky question.

LIGHT (AND DARKNESS) ON THE HORIZON

There's another side to the coin, however. If you look at the big picture, there are some remarkably positive aspects, and I'd like to suggest that, far from being ecological vandals, astronomers and space scientists can be advocates for good environmental management. Of course, you could argue that I *would* say that, being one myself—but I hope you will agree with me by the end of the chapter.

To start with, you need look no further than the promotion by optical astronomers of environmentally friendly outdoor lighting. While astronomers have no wish to dim the streets of the world's cities into depressing twilight zones, they do advocate good lighting design as a means of preventing ground-based astronomy's eventual demise.

Information from the heavens comes to us primarily in the form of various flavours of radiation forming what is known as the electromagnetic spectrum. It includes gamma rays, X-rays and radio waves as well as visible, ultraviolet and infrared light. We can also detect subatomic particles from the Universe with specialised telescopes, and, eventually, we will have access to the gravitational waves that ripple through the very fabric of space-time. But electromagnetic astronomy is still the main game. Astronomers use every part of the electromagnetic spectrum to make their observations but have to rely on space-borne or balloon-borne telescopes for all but visible-light, infrared and radio work. Scientists using these particular wavebands have the luxury of being able to make their observations from the ground. But all ground-based astronomy—including radio astronomy—is susceptible to the detrimental effects of human-made interference. And ground-based optical telescopes do need dark skies.

Something that surprises most visitors to an observatory site is not how dark the night sky is but how much light actually comes from it. When there's no Moon in the sky, the absence of light pollution renders the stars dazzling. A closer look, however, reveals that it's not just the stars that are bright: the sky itself is luminous, a phenomenon made more obvious by the presence of clouds, which appear jet black against their background. They contrast strikingly with their counterparts in a brightly illuminated city sky. What causes this natural sky glow? The brightest component is barcode-type emissions generated by atoms in the upper atmosphere relaxing after a hard day in the Sun. Oxygen, nitrogen and something known as the 'hydroxyl molecule' are key contributors to this. In addition, there is a rainbow spectrum background whose sources include illuminated dust in the Solar System, faint stars within our own Milky Way Galaxy and, at the faintest level, the light of distant galaxies that are indistinguishable from one another.

Any celestial object that an astronomer wishes to investigate is superimposed on this sky background, and usually the background dominates the observation. Often, astronomers are looking for faint objects whose brightness is less than 1 per cent of the natural sky background's brightness. Little wonder that any increase in the background glow due to artificial light renders these objects invisible. In that respect, there are few natural environments as sensitive to pollution as the night sky.

In the 1980s, astrophysicists Roy Garstang and David Crawford did pioneering work in the United States on the spread of urban sky glow, with the aim of meeting the challenge of keeping observatory sites dark. That led to the formation of the International Dark Sky Association, which is today the world's premier advocacy body for good outdoor lighting. In the early 2000s, these studies were taken further by astronomer Pierantonio Cinzano, at the

University of Padua in Italy. Cinzano mapped the world's light pollution using night-time images from space (to show the upward-pointing light sources) combined with the properties of the atmosphere in spreading light around. His calculations are highly refined, taking into account the curvature of the Earth, and show that two-thirds of the world's population lives in light-polluted conditions. More dramatically, about one-fifth of the population can no longer see the arc of the Milky Way.

Cinzano's studies also show that light pollution is starting to interfere with many of the world's major observatory sites. It seems that that remoteness is not enough to protect an observatory, and steps have to be taken to reduce the spread of light pollution. They can be legislative or educational, but they need to bring home the message that light pollution is bad news for everyone. Not only are skywatchers and nocturnal animal species directly affected, but the disruptive effects of obtrusive lighting on our own circadian rhythms are now starting to be recognised. And, of course, the carbon footprint of light shining where it's not needed is of universal concern.

A number of the world's nations have begun to protect their night skies with legislation. They include the United States, the United Kingdom, Australia and parts of Italy. In the Czech Republic, standards introduced in 2002 defined light pollution as 'every form of illumination by artificial light that is dispersed outside the areas it is dedicated to, particularly if directed above the level of the horizon'. The ruling came about because of popular concern about the glare caused by poorly designed outdoor lighting. Fully shielded fixtures allowing only downward illumination—the favourites of astronomers—are the preferred option in the Czech Republic, as elsewhere.

Meanwhile, Siding Spring Mountain, in north-western New South Wales, is home to Australia's national optical observatory, which provides Australian optical astronomers with their largest observing facility on domestic soil—the AAT. It is also home to several smaller telescopes and, because it represents a major infrastructure investment, is protected by planning regulations extending to a distance of more than 100 kilometres from the site. In the central part of this zone, fully shielded (or 'full cut-off') outdoor light fittings are required. Beyond the regulated zone, though, there is no legislation to protect the skies of the observatory—even though the night-time glow of the Sydney–Newcastle metropolitan area (at a distance of about 350 kilometres) can clearly be seen, as predicted by Cinzano's model. The astronomers of Siding Spring Observatory take a pragmatic view of this problem. They don't necessarily want to see lights switched off, but they do want them better designed. For those astronomers, it's a question of winning hearts and minds—particularly outside the regulated area—and the developers, makers and sellers of outdoor lighting fixtures are their prime targets. It is no exaggeration to say that, in some degree at least, these folk hold the future of Australian optical astronomy in their hands.

THE UNGREENING OF MARS

The idea of astronomers becoming activists in the fight against light pollution has no real equivalent in the space world. Of course, there are strident voices within the industry demanding a clean-up of the Earth's environment—it's in their own interests to do so. Likewise, both inside and outside Russia there are outspoken critics of the ongoing use of toxic propellants in launch vehicles like Proton, but there is no sign of an end to this technology. Rather, one has to look at the outcomes of space exploration, and here there are definite positives for the environment. Pierantonio Cinzano's work using space images of the night-time Earth is just one example. The entire field of remote sensing would not exist had it not been for space exploration, and the Earth is much better for it. Land use surveys, logging surveillance, ocean temperature surveys, ice-cap melt rates—the list is almost endless and growing daily. We live in an age in which we know our planet better than at any other time, thanks to orbiting spacecraft—and we all benefit from it.

It's worth looking further afield, however, to seek other evidence of space science's contribution to the Earth's well-being. One of our nearest neighbours, Mars, is helping us to understand our own planet's behaviour. Mars is about half the size of the Earth, and about 50 per cent further from the Sun than we are. Of all the Solar System's planets and moons, it is the world most similar to our own, having a rocky surface and predominantly clear skies in a dust-dry atmosphere that has about 1 per cent of the pressure of ours. Being further from the Sun, Mars has a lower surface temperature than the Earth: near the equator, it typically ranges between about –78 and –8 degrees Celsius—although higher and lower extremes are possible. By coincidence, its orbital tilt is similar to Earth's, and its day is only a little different, at 24 hours and 40 minutes.

The exploration of Mars using robotic spacecraft has reached epic proportions since the first *Viking* landers successfully touched down on the planet's surface, in 1976. In mid-2012, there were three active satellites orbiting Mars (*Mars Odyssey*, *Mars Express* and *Mars Reconnaisance Orbiter*) and one operational NASA rover on the surface (*Opportunity*, whose extraordinarily successful twin, *Spirit*, became defunct in 2010). But on 6 August 2012, a new, car-sized rover, *Curiosity*, landed on the planet's surface amid a deluge of popular interest almost unprecedented since the *Apollo* Moon landings. *Curiosity*'s task is to use its mobile laboratory to probe the red planet's geology, searching for evidence of past habitability.

Though its surface is now dry and dusty, Mars is known to have considerable deposits of water ice both on and beneath its surface. Extensive polar ice-caps have been known since the first telescopic studies, but orbital imaging has revealed ice patches around several craters, an ice lake in Mars' northern plains and what looks like dust-covered pack ice covering an area the size of New South Wales close to the equator. When NASA's *Phoenix* lander sampled the northern arctic in the Martian summer of 2008, it found an extensive layer of ice a few centimetres below the surface soil. The spacecraft also detected the presence of a natural antifreeze in the form of mineral salts capable of lowering the freezing point of water by as much as 70 degrees Celsius—enough to allow pockets of liquid water to exist on the surface.

Meanwhile, the robotic exploration of Mars has provided increasing evidence of an ancient ocean having once occupied much of its northern hemisphere. Both the mineral composition of the surface and its physical features suggest the former existence of large bodies of water, with well-defined coastlines, estuaries, beaches and sea cliffs. The ocean is thought to have existed 3.8 billion years ago, when the planet was only 800 million years old. We don't yet know how long it lasted, but it would have required a thicker atmosphere than today's to allow the greenhouse effect to sustain a similar average temperature to that of the modern Earth. Whether this atmosphere was formed with the planet or resulted from later events such as asteroid impacts is still in doubt. Either way, something happened to change things. A clue comes from Mars' almost complete lack of magnetic field, suggesting the absence of a liquid iron core. The interior of the planet is probably much cooler than the Earth's and not hot enough to allow the movement of continental plates (although there is evidence of past tectonic activity, including the Solar System's tallest volcano).

We know that on Earth there is a regulated cycle in which carbon is exchanged between the atmosphere and the planet's rocky crust. Volcanoes feed carbon into the atmosphere, which in turn finds its way into the oceans via rainfall, eventually falling to the ocean floor to be incorporated into the mantle by the incessant motion of the continental plates. The cycle forms a kind of thermostat, keeping just the right amount of carbon in the atmosphere to provide a greenhouse blanket around the planet. In the case of Mars, this seems to have shut down early in the planet's history. Its greenhouse thermostat would have ceased to function, and Mars' blanket of atmospheric carbon would have dissipated, allowing the surface temperature to fall to its present frigid levels. A proportion of the Martian ocean's water would have evaporated into space, having been broken up into its component atoms, while the rest would have been frozen into the ice deposits we see today. The real culprit for the loss of Mars' ocean is not the planet's greater distance from the Sun compared with Earth's, but its smaller size. It seems that Mars was simply too small to sustain the tectonic processes that regulate our own atmosphere.

Our understanding of the processes going on in the Earth is enormously enhanced by our ability to study an additional, and quite different, example of planetary geophysics in the shape of our diminutive neighbour. And then there's Venus—which went the other way with a runaway greenhouse effect, and now has a surface temperature hot enough to melt lead. But that is a story for another time.

THE VIEW FROM ABOVE

There is one further aspect of space flight that I think is worth mentioning in the context of environmental awareness, and which is particularly relevant in a book set against a backdrop of travel and tourism. Unlikely as it may seem, the fledgling space tourism industry may have something to offer in improving our planet's well-being.

At present, we are on the brink of a revolution in humankind's access to space. The 450 or so individuals who have experienced space flight first hand have done so almost entirely under the auspices of government-sponsored exploration. The few exceptions are a handful of unbelievably wealthy folk who have taken advantage of unoccupied seats on *Soyuz* supply missions to the International Space Station in a $150 million commercial deal brokered with the Russian Space Agency. One of the principal players in this orbital wheeler-dealing, incidentally, is an Aussie.

But a few other visionary entrepreneurs want to change the rules and bring the experience of space to a much wider clientele. The idea is to provide a suborbital flight, which will carry passengers to the edge of space, where they can see the curvature of the Earth and experience the phenomenon of weightlessness. They will look down on the atmosphere from the blackness of space. The flights will be made in rocket planes carrying about half-a-dozen passengers at a time, with a simple ascent–descent flight profile. Typically, a 90-second rocket burn will result in a maximum speed of 1.5 kilometres per second. As we have seen, this is well below the threshold for orbital flight, but it is sufficient to hurl the craft up to a height of 100 kilometres after the motor has shut down, allowing three or four minutes of weightless free fall before braking commences.

Leading the charge is Richard Branson's Virgin Galactic, whose *Enterprise* rocket plane is the first of five planned vehicles in Virgin's space fleet. It will commence each flight slung under the wing of an unconventional jet aircraft—the 'mother ship'. Taking off from Virgin's Spaceport America at Upham (yes, truly!), New Mexico, *Enterprise* will be ferried to a height of 15 kilometres before being released and pitched upwards to make its dramatic powered ascent. Having reached apogee—its most distant point from the Earth's centre—the rocket plane will then fall back towards Earth, turning into a glider to land in much the same manner as a space shuttle. Virgin Galactic's development program has progressed well, and flights with fare-paying passengers—of whom there are already several hundred on the waiting list—are expected to begin soon after this book appears in print.

Despite the fact that the fare is US$200 000, Virgin Galactic is confident of its market for such flights and expects the price to fall as the project progresses. Safety is paramount—nothing would damage the infant venture more than the loss of a rocket plane and its passengers. On the other hand, over-regulation could stifle progress of this next step in tourism, so legislators face a delicate balance in exercising control—one not seen, perhaps, since the early days of aviation.

Before you raise your hands in horror at the thought of well-heeled joy-riders hooning into space—and to hell with the atmosphere—there are a couple of additional points to consider. First, despite the relative toxicity of *Enterprise*'s hybrid propellant mix, its total carbon footprint per passenger per flight is less than half that of a one-way trans-Atlantic flight. And, initially, at least, space rides will be small in number. By the time space tourism becomes big enough to rate as a significant atmospheric polluter, the technology is likely to have improved considerably in its green credentials.

Perhaps even more significant, though, is the sentiment that will be evoked among tomorrow's space tourists. We have all felt something of it ourselves in viewing images of the Earth from space, whether it be the famous *Earthrise* image from *Apollo 8* in 1968, or recent snapshots from the Cupola—the 'bay window' of the International Space Station. We cannot help but marvel at the fragility of the environment that sustains the biosphere. That thin blue membrane of air visible from space—a vanishingly small envelope around what is itself only a modestly sized planet—is compelling in its message of the frailty of our surroundings. Seen at first hand from space, this can only underline the true vulnerable nature of the world in which we live. Indeed, many astronauts have commented that they have gained a profound regard for the environment after their experience of space flight. It's rather like the sense of awe that the Kimberley coast inspired among those of us aboard *Orion* at the beginning of the chapter—only much more intense. And that is my hope. For the first few years, at least, space tourism will be the province of the rich and influential. If these people return to Earth with a profoundly enhanced respect for their planet, perhaps the message will finally get through to the rest of us.

10 DARK SECRETS

Astronomy's big questions

During the 1980s, a most unusual astronomy book was published. Written by an astronomer working with a designer of folded-paper models, it was an ambitious and elaborate 'pop-up' book of the Universe. The book made a heroic effort to represent the wonders of the Cosmos in three dimensions, and I think it had its moment in the bookshops. But, with all due respect to the authors (one of whom I count among my friends), it has to be said that the wonders of the Cosmos really don't lend themselves to folded-paper models—certainly not as successfully as the usual stock-in-trade of the pop-up genre: Jemima Puddleduck, Postman Pat and Thomas the Tank Engine. When it comes to planets, stars and galaxies, the medium becomes a little inadequate. To be honest, a folded-paper planet looks more like a hibernating armadillo. A star in the process of formation resembles something you'd hang on a Christmas tree. And a galaxy looks strikingly like the aftermath of some bizarre culinary experiment involving a soufflé and a stick of dynamite. Of all the pop-ups in the book, though, the least awe-inspiring is the pop-up Big Bang. The cataclysmic explosion that gave birth to the entire Universe is reduced to a series of creaks and shuffles as you open the book. No bang. Not even a pop. Just a garish paper splodge lurching unsteadily into existence before your eyes.

'Well, what do you expect?' I can almost hear the indignant author saying over my shoulder. And she would have a point. What *do* we imagine the Big Bang to have been like? Nothing, absolutely nothing in our experience allows us to envisage even remotely the power of this most significant event in the Universe's history. We can imagine explosions, of course. Even nuclear ones. But in our mind's eye we always see them from the outside. And the trouble with the Big Bang is that when it went off, 13.7 billion years ago, it not only produced everything now contained in the Universe but created space and probably time as well. There was no 'outside' for the infant Universe to explode into—all of space was contained within its violently expanding boundaries. If, indeed, there were any boundaries. And there was no 'before', if time itself started with the instant of creation. These are truly weird concepts. Perhaps, after all, the pop-up book did make a pretty good attempt at depicting the undepictable.

PUTTING THE BANG INTO THE UNIVERSE

The supremely understated name we give to this extreme event came from someone who was actually sceptical about it. In 1949, on BBC radio, Fred Hoyle (who was then merely a professor rather than a sir) made a disparaging remark about the theory, using the term Big Bang in an attempt to highlight how ridiculous it was. Once his radio lectures were enshrined in print in a little book called *The Nature of the Universe*, the name stuck, and we've been stuck with it ever since. You'd have thought that cosmologists could have come up with something far more elegant to describe the single most important event in the history of the Universe, but I'm afraid they haven't.

In fact, the theory itself has a much more imposing pedigree than its name suggests. It has its origins in the aftermath of Albert Einstein's General Theory of Relativity, which, I'm sure you'll remember, was published early in 1916. Little more than a year later, however, Einstein thought his new theory was in big trouble. If his mathematical equations were applied to the Universe as a whole, they became unstable, producing a Universe that would be changing in size. To the best of Einstein's knowledge, the Universe was static. So he did something clever: he introduced a mathematical entity that he called the 'cosmological constant'. It could have a positive or negative effect and would represent an inbuilt attractive or repulsive force in the fabric of space that would balance any tendency for it to expand or contract. As far as Einstein was concerned, that solved the problem. His equations then represented a well-behaved static Universe—and he could sleep easy at night. Phew.

But in 1912 evidence had already begun to emerge that the Universe might not be static at all. In that year, Vesto Melvin Slipher of the Lowell Observatory (from where Pluto was later discovered) had embarked on the first systematic measurement of the radial velocities of the mysterious objects then known as 'spiral nebulae'. Perhaps you'll recall from Chapter 6 that such an undertaking required the use of a spectrograph and today is called a 'galaxy redshift survey'. Slipher's work, trivial though it may seem to us now, was a triumph of observational astronomy. Each of his objects required between 20 and 40 hours of photographic exposure time on the Lowell 0.6-metre refracting telescope, gathered over several nights, producing spectra whose barcode features were even then barely distinguishable. This was no mean feat, for, only a few years earlier, the great 1.5-metre reflecting telescope at Mount Wilson Observatory in California had needed no fewer than 80 hours to obtain a single spectrum of one of these objects. Slipher's results, published in 1917, showed that the 25 objects he had managed to observe were predominantly receding rather than approaching. The spiral nebulae—whatever they were—were racing away from us.

When, eight years later, Edwin Hubble showed conclusively that the nebulae were galaxies—huge, remote objects rather than nearby, small ones—Slipher's results took on new significance. They were the first hint of a relationship between the velocity of a galaxy and its distance from our own Milky Way. The further away a galaxy was, the faster it was racing away from us. And it was Hubble again, working at Mount Wilson with the new 2.5-metre telescope (then the biggest in the world), who spectacularly confirmed the relationship, in 1929. Which, of course, is why it's now called Hubble's Law.

Meanwhile, two other mathematicians, Willem de Sitter (the Dutchman we met in Chapter 8) and a Russian scientist called Alexander Friedman, had produced new solutions to Einstein's equations that rashly allowed for an expanding Universe. By 1927, a Belgian priest called Georges Lemaître had taken their work further by proposing a Universe in which distant galaxies would appear to recede from the observer at greater speeds than those nearby, thus foreshadowing Hubble's Law itself. Following Hubble's confirmation of the relationship, its implication became clear—that the Universe was expanding everywhere at a constant rate, neither accelerating nor decelerating in its expansion. Lemaître then took the next big step, reasoning that if the Universe was expanding at a constant rate there must have been a time in the distant past when everything was in the same place. Thus, in 1931, Lemaître produced his theory of a primordial atom from which everything had expanded—the forerunner of today's Big Bang model.

Much of the refinement leading to the present version of the model was introduced in the 1940s by a Ukrainian-born US physicist called George Gamow. It included the idea that the early Universe had an extremely high density and temperature, and suggested that this would have led to the formation of hydrogen and helium at levels matching those that are observed today. The prediction was made in a paper jointly authored with physicists Ralph Asher Alpher and Hans Albrecht Bethe—both of whom played a significantly greater role than merely lending their names for the celebrated pun on the first three letters of the Greek alphabet. The 'Alpher, Bethe, Gamow' paper was published on 1 April 1948 and, despite the date, contained ground-breaking research.

Gamow also developed the extraordinary idea that if we could see far enough back in time, by looking deep into space, we would be able to see the flash of the Big Bang itself. We would be able to see back to a time just before the infant Universe became transparent, when it was still filled with a brightly glowing fog of radiation. Were it not for the expansion of the Universe since the light was emitted—a consequence of the Big Bang itself—we would expect to see this as visible light. The entire sky would shine brightly, rendering the stars invisible, and we would know little about our place in space. But as the Universe has expanded, the light travelling through it will have been stretched into faint whispers of microwave radiation.

Eventually, in 1964, that radiation was famously discovered—completely by accident—when a new radio astronomy antenna was being tested in Holmdel, New Jersey. Scientists Arno Allan Penzias and Robert Wilson tried everything they could to eliminate a mysterious background signal that seemed to emanate from the entire sky. They even cleared pigeon droppings from the antenna. (Doesn't that conjure up the most bizarre conversation? 'Oh, it's the flash of the Big Bang. We thought it was just pigeon poo . . .') The supposedly faulty signal is now known as the 'cosmic microwave background radiation' (CMBR), and it is the most ancient thing we can ever see. Dating from a time only about 380 000 years after the Big Bang itself, this fossil radiation has been travelling through space for almost the complete lifetime of the Universe.

The CMBR was the clinching proof of the Big Bang model. It was also the final nail in the coffin for its main rival, the Steady State Theory of the Universe, which had been championed by Fred Hoyle, among others. Happily, both Georges Lemaître and George Gamow were still alive to see the discovery of the CMBR. And happily for Penzias and Wilson, in 1978, they jointly received the Nobel Prize for Physics for their work.

But what of Einstein and the cosmological constant he'd introduced to make the Universe nicely static? Once the expansion had been confirmed by Hubble in 1929, Einstein quickly withdrew his idea in embarrassment.

And, years afterwards, George Gamow disclosed that, 'When I was discussing cosmological problems with Einstein, he remarked that the introduction of the cosmological term was the biggest blunder he ever made in his life.' However, the latest findings in cosmology suggest that, far from blundering, Einstein had shown great insight in introducing the cosmological constant. Remarkably, as we will see later in this chapter, an inbuilt repulsive force in the fabric of space truly seems to be there. Perhaps less remarkably, Einstein himself was an even bigger genius than he realised.

The fascination of the public at large with these discoveries about the wider Universe has, over the years, fed a growing industry in popular cosmology. The hapless pop-up book is but one recent example, but the genre goes back a long way. Almost before the ink was dry on Einstein's general relativity paper, he had produced a popular-level book on the subject, *Relativity: The Special and the General Theory*, which was first published in 1916. Many more scientists, from Eddington to Hoyle and beyond, produced accessible accounts, and who could forget Gamow's fictional hero, Mr Tompkins, a humble bank clerk who married the daughter of a physics professor? Dreaming his way through the byways of modern physics, Mr Tompkins brought science to the people during the dark days of the Second World War.

If Mr Tompkins was an unlikely exponent of the mysteries of the Universe, an even more unlikely one turned up in the 1990s, in the shape of a British stand-up comic. In a three-part TV series made for the United Kingdom's Channel 4, the late Ken Campbell took the role of an ordinary bloke driven by curiosity to probe the secrets of the Universe on both cosmic and quantum scales—the very large and the very small. In *Reality on the Rocks*, Campbell's visit to the 4.2-metre William Herschel Telescope, on the island of La Palma in the Canary Islands, vied with his exploration of the giant particle accelerator at the European Centre for Nuclear Research, in Geneva, for spectacular footage. But *Reality on the Rocks* was no run-of-the mill science doco. The whizz-bang stuff was cut with candid scenes from Campbell's one-man theatre show, producing a delicious mix of enlightenment, hard-hitting comedy and a quest for meaning. Forget the movie star voice-overs and deep, meaningful music that accompany most astronomy documentaries—this was science in the raw.

I mention *Reality on the Rocks* not just because I thought it did a brilliant job in popularising physics, or because I happened to stumble into it as a minor player (courtesy of an observing stint on the William Herschel Telescope), but also because it endowed me with a lasting sense of optimism about our growing understanding of the Universe. Campbell's dogged determination to comprehend what we know about the Universe and place it into the perspective of everyday life was inspiring in itself—especially when seen at close quarters. However, it was a humble Spanish café owner in the small coastal town of Tazacorte, on La Palma, who unwittingly hit the nail on the head. I can't be sure, but I think his name was José. We were shooting a final scene around the tables outside this gentleman's restaurant on the fringe of Tazacorte's black-sand beach, chatting about the Universe and enjoying the seafood delicacies that he insisted on giving us. With the setting Sun blazing over the Atlantic Ocean in front of us, it was the classic TV atmosphere shot.

Eventually, someone asked José what *he* thought about the Universe. 'Ah, yes,' he replied, sagely. 'Yes. Things have been better since the Universe was here. Much better.'

I think it's the single most profound thing I've ever heard. OK, his command of English didn't allow him to differentiate between the words 'universe' and 'observatory', but what a way to sum up all the motivation and aspiration of astronomers and cosmologists. I doubt his remark made the final cut—I'm afraid I can't remember—but that thoughtful man's words have stuck with me ever since.

GROPING IN THE DARK

When *Reality on the Rocks* went to air in 1995, science's view of the Universe as a whole was troubled with inconsistencies. A couple of them had been fixed a decade or so earlier, when the concept of inflation was introduced, which says that the Universe expanded by a colossal amount during the first gazillionth of a second of its existence—from the size of a subatomic particle to the size of a galaxy within 10^{-33} of a second of the Big Bang. (That's a 1 with a decimal point and 32 zeroes in front of it.) Then, for some reason, inflation stopped. While financial inflation is usually a bad thing, cosmic inflation solved problems of geometry (the absence of large-scale curvature of space-time) and uniformity (the smoothness of the CMBR) in our understanding of the early Universe. That modification of the theory still holds good today, and most scientists accept the inflationary model of the Big Bang.

But another geometry problem niggled at the fertile minds of cosmologists. When you added up all the matter in the Cosmos, it looked as if there wasn't enough to give the Universe its observed geometry. The conventional wisdom was that the mutual gravitational pull of all the matter in the Universe—galaxies, stars, planets and all the rest—should conspire to slow down its expansion over time. In other words, the Universe should be decelerating, and this slow-down should be detectable by looking far enough into space to find a departure from Hubble's Law. It should produce a particular geometric signature for space on the largest scale. But that didn't seem to be there, and one or two scientists wondered whether something was badly wrong with the theory.

Those near-heretics were vindicated dramatically in 1998, when a group of astronomers led by Brian P. Schmidt of the Australian National University produced hard evidence that, far from slowing down as expected, the Universe is expanding more rapidly today than it was seven or eight billion years ago. Bizarrely, the expansion of the Universe is accelerating. Confirmation came from a rival group within the year, and the discovery was hailed by *Science* magazine as the 'breakthrough of the year' for 1998. It's a great pleasure to report that the accolades have not subsided, as, a few weeks before these words were written, Schmidt and his colleague, Adam G. Riess, together with Saul Perlmutter—leader of the rival team in the United States—were awarded the 2011 Nobel Prize for Physics. The partying throughout Australian astronomy seems set to go on for quite a while yet, and Schmidt even got a mention by the Queen during her visit shortly afterwards. Wow.

The evidence for accelerating expansion collected by these two groups came in the form of observations of a particular kind of supernova—the Type Ia—at very great distances. Supernovae of this type are caused by old stars exploding violently as a result of matter being deposited onto them from a nearby companion star. They provide extremely bright standard candles, easily outshining their host galaxies. What caused all the excitement was that these remote supernovae were dimmer than they ought to have been, given their estimated distances—and hence look-back times—from our own Galaxy. This suggests that, yes, the expansion of the Universe is accelerating, and that has now been confirmed by a number of different methods.

The effect is attributed to an inherent springiness of space—or dark energy—that is overcoming the tendency of the Universe to decelerate because of the mutual gravitational attraction of everything within it. It's described as a 'negative pressure' to distinguish it from the positive pressure experienced at the centre of a cloud of gas collapsing under its own gravity, for example. In other words, it's a tension. Moreover, we now know that this dark energy is the largest single component of the Universe, amounting to 72 per cent of its total mass/energy budget. (I'm sure you remember that mass and energy are equivalent, by courtesy of Einstein's special relativity.)

So what exactly is dark energy? When consideration was given to this question, an intriguing possibility emerged. After Einstein's embarrassing withdrawal of the concept of a cosmological constant, most scientists had simply assumed that the constant was zero and that space had no inbuilt force field. But could the newly discovered dark energy be something to do with this long-neglected orphan of general relativity? If it had the form of a constant negative pressure, and the pressure was so weak that it only began to overcome gravity when the characteristic distances separating galaxies had become very large—a long time after the Big Bang—then it might just fit the bill.

There are other theoretical possibilities, too, but they require the introduction of various flavours of 'new physics'—those underlying realities, such as quantum gravity and string theory, that are not predicted by relativity. These are well beyond the current limits of certainty, and are active areas of research. Often in such investigations, scientists end up following blind alleys, and the certainties only emerge over time. One such possibility is a new fundamental force—perhaps the whimsically named 'quintessence'. Its relationship to the four fundamental forces we already know about echoes the fifth element of ancient Greek philosophy, the heavenly substance of the stars. Like the effect of the cosmological constant, this would have to be a dark energy with negative pressure, but one key difference is that quintessence would change with time, leaving its imprint on the Universe only in the relatively recent past.

Another possibility that must be explored is to abandon the hallowed cosmological principle, which says the Universe is the same in all directions. That would permit a Universe that has significant differences between one place and another, perhaps again eliminating the need for an overall repulsive pressure. This possibility may have been ruled out by recent work at Johns Hopkins University, in Baltimore, which has examined the idea of a local void in the Universe giving the illusion of accelerated expansion. It appears that such a hypothetical void cannot be consistent with the most recent measurements of the expansion velocity.

Most astronomers today accept the reality of dark energy—and its consequences for the long-term future of the Universe. When, during the 1970s and 1980s, we believed that the Universe's expansion was slowing down, many astronomers thought the expansion would eventually reverse, turning into a collapse that would culminate in what was usually called the 'Big Crunch'. It would be a rewind of the Big Bang, in a sense, which is why Brian Schmidt always calls it the 'gnaB giB'. Yes, quite so. But with the discovery of the accelerating expansion, that scenario changed, and we now expect the Universe to continue expanding forever.

Sadly, the Universe seems destined to become an incredibly boring place as a result. Its reserves of hydrogen will be consumed in stars, which are themselves doomed to die when their nuclear fuel runs out. Moreover, the accelerating expansion will eventually carry most galaxies beyond the horizon of visibility, because they will be receding faster than the speed of light. Any intelligent beings will have no idea that there are other galaxies, or that the Universe started in a Big Bang. It will make the science of cosmology very difficult indeed.

Some of today's scientists—including Schmidt—see an even more startling future. If the expansion continues to accelerate, it's possible that space itself may be torn apart in a scenario that has been dubbed the 'Big Rip'. It's very hard for us to imagine this, and even experts in the field disagree about the possibility because, as yet, we have no accepted theory of exactly how empty space is constructed. Still, it makes for good headlines.

The issue with dark energy is that no one really knows what causes it. The quantum physicists think it might be the result of a seething foam of virtual particles popping in and out of existence and imbuing the fabric of space with a negative pressure. It sounds plausible, except that even the best of these theories predicts a repulsive force that is 10^{120} times bigger than what we observe (yes, that's a 1 followed by 120 zeros). A repulsive force as intense as that would already have made its presence felt by tearing everything apart—including atoms—in a baby-Universe version of the Big Rip. No wonder this estimate is considered by most physicists to be the worst theoretical prediction in the whole of science.

What can we do to improve our flawed understanding of the problem? A good start would be to identify which model of dark energy best fits the astronomical observations—cosmological constant, quintessence or something else? Not surprisingly, the required observations are hard to carry out, but they are underway. A number of groups throughout the world are now actively engaged in tackling the dark energy problem, typically by extending the standard candle supernova observations to greater distances and greater numbers of objects.

There's another possibility, however, based on the fact that dark energy has a significant influence on the large-scale geometry of the Universe. If that geometry can be probed at different periods in the Universe's history, perhaps by using some kind of standard ruler seen at varying distances (for example, the typical separation of pairs of galaxies, which seems to be the same throughout the Universe), there is a real chance that the correct model of dark energy will be identified. We have a good starting point in this quest, through our knowledge of the large-scale geometry at two key times in cosmic history. One is the very early Universe. That microwave background radiation we mentioned a couple of pages ago, the CMBR, is a kind of cosmic wallpaper pasted across the entire sky and is behind (or, more accurately, *before*) everything else in the visible Universe. Early studies showed it had a temperature of 2.7 degrees Kelvin (i.e. degrees above absolute zero), which is about what you would expect from 13.7 billion years of cooling off. But more detailed investigations revealed minute temperature variations in the radiation from one part of the sky to another at the minuscule level of 0.00001 degrees Kelvin. These had their origin in acoustic oscillations—sound-waves—in the primordial fireball. They represent the *bang* of the Big Bang frozen in time, and they give us an accurate picture of what the Universe was like in its infancy. And then, as we saw in Chapter 6, today's Universe has also been thoroughly explored with large-scale surveys of the three-dimensional distribution of galaxies, such as the 2dF Galaxy Redshift Survey, made with the 3.9-metre AAT. Both these snapshots of the Universe reveal structure that is of great interest to cosmologists. Those minute variations in the CMBR have been explored in ever-finer detail by a succession of spacecraft, beginning with the *Cosmic Background Explorer*, in the early 1990s, followed by the *Wilkinson Microwave Anisotropy Probe*, in the early 2000s, and culminating in today's *Planck* spacecraft, whose findings will be announced as this book goes to press. And the redshift surveys tell us that the distribution of galaxies in today's Universe is spidery, resembling a honeycomb or foam of galaxies. In a remarkable vindication of our understanding of how the Universe works, this honeycomb structure is exactly what you would expect to see if you could fast-forward the patterns in the CMBR to the present time.

Comparison of the Universe's geometry at these two periods has already allowed a wealth of information on its evolution to be deduced. But the missing ingredient has been a similar three-dimensional survey of galaxies at great enough distances that they correspond to a look-back time of about half the age of the Universe—a time seven or eight billion years ago when dark energy first began to make its presence felt. Investigating the standard rulers in such a survey would require observations of hundreds of thousands of faint galaxies, a hugely ambitious program. That has now been at least partially provided by a number of galaxy surveys, including one named WiggleZ, which was carried out on the AAT. The Z in WiggleZ is the scientific symbol for redshift, but the Wiggle part is all about the acoustic oscillations and the structure they imprinted as the Universe evolved. Completed in January 2011, WiggleZ measured the redshifts of 200 000 galaxies, mapping the cosmic structure across look-back times of up to eight billion years.

Further down the track, studies such as this could be carried out with even greater precision if bigger telescopes were used, allowing fainter galaxies to be observed. That will be the province of the new generation of 20- to 30-metre telescopes—the extremely large telescopes. These represent the latest step in the evolution of the telescope and are no more than a decade away—at least in terms of their construction. There are several projects under consideration, including E-ELT (the 39-metre European Extremely Large Telescope), the US Thirty Meter Telescope and the international Giant Magellan Telescope, in which Australia is already a partner. The last of these is of particular interest for studies of dark energy, as it will have a sufficiently wide field of view to make it a formidable survey telescope rather than simply an instrument that studies single objects in minute detail.

So, what results are we acquiring from studies such as WiggleZ and distant supernova measurements? Both these techniques have returned information on the nature of dark energy that has a decided preference for Einstein's old nemesis, the cosmological constant. Supernova studies indicate that the negative pressure of dark energy seems to have changed by less than 20 per cent since the Universe was about half its current size. Likewise, the independent WiggleZ measurements are best fit by a model with a constant negative pressure. These are truly remarkable findings, not least because they strongly hint that even when Einstein thought he was blundering, he was actually right. You really do have to take your hat off to him.

WONDERING WHAT'S THE MATTER?

As you might expect, astronomers are pretty frustrated that they don't know what 72 per cent of the Universe is made of. It keeps some of them awake at night. But they do know what the other 28 per cent is, right?

Wrong, I'm afraid. Just when you thought you were getting your head around the mass/energy budget of the Universe, along comes something else to thwart you. In fact, the other big unknown—dark matter—has a much longer history in scientific research than dark energy. We've been mystified for longer, but, on the positive side, we could be closer to finding out what it is.

How do we know dark matter is present in the Universe? It reveals itself only by one thing—the effect of its gravitational attraction on matter that we can see. Other than that, we have no way of detecting it, at least for the time being. We can, however, sense this gravitational smoking gun by a number of means—described below—and they all give the same answer: that the ordinary matter we see by the radiation it emits (for example, in stars and glowing gas clouds) or absorbs (for example, in dust clouds silhouetted against a bright background) amounts to only one-sixth of what gravity tells us is there. Embarrassingly, dark matter outweighs visible matter by five to one.

In fact, the very first estimate of how much dark matter there is gave an even higher imbalance, because at the time of its discovery astronomers didn't know about the copious quantities of perfectly normal matter in the form of hydrogen that surrounds most galaxies. The man who first noticed that something didn't add up—and thereby stumbled across dark matter—was one of the great characters of twentieth-century science. He was a Swiss-US astronomer by the name of Fritz Zwicky, and he was interested in clusters of galaxies, the largest concentrations of matter in the Universe. Like many of his contemporaries, Zwicky didn't suffer fools gladly, and once famously described some of his colleagues not merely as 'bastards' but as 'spherical bastards'. Why? Because, according to Zwicky, they were bastards whichever way you looked at them—and the only thing that looks the same from all directions is a sphere. He did have a way with words.

In 1933, the 35-year-old Zwicky was studying a cluster of galaxies in the constellation of Coma Berenices (Berenice's Hair) in the northern-hemisphere sky. With the tried and tested method of spectroscopy, he used the Doppler effect to measure the radial velocities of several members of the cluster. He was astonished to find that these galaxies seemed to be moving too quickly, relative to the cluster, for its gravity to hold on to them. Given their velocities, the galaxies he was observing should have escaped long ago, because the gravitational attraction of all the visible matter was simply not enough to stop the cluster disintegrating. Zwicky calculated that it would require 400 times more mass than he could account for to keep the cluster intact—an over-estimate, as I have mentioned, but an understandable one. However, he was spot on in his inference that something invisible was present, a component that neither emitted light nor absorbed it from the radiation of background objects.

Surprisingly, not a lot happened as a result of Zwicky's observations, since astronomers didn't really understand them. It was not until 1970 that a lone voice was raised in concern about the behaviour of galaxies themselves. That voice belonged to a 30-year-old Australian researcher by the name of Kenneth C. Freeman—who had set about investigating the way galaxies rotate, and had discovered new evidence for the existence of dark matter. Once again, the underlying principle was to measure the motion of objects using the Doppler effect, but this time the objects were not galaxies in clusters, but clouds of gas in individual galaxies.

The trick here is to measure the characteristic speeds of objects within a galaxy—stars or gas clouds, it doesn't really matter—and look at the way those speeds change between objects moving around the centre and those moving around the edge. The result is a graph called a 'rotation curve', and it is easiest to measure in spiral galaxies that are almost edge on to our line of sight. Basic orbital dynamics tells you that if you assume the mass of your galaxy is concentrated where it is brightest, then objects closest to this point will whizz around much more rapidly than those further out. Thus, you would expect the rotation curve to be highest near the centre and steadily decline towards the edge—exactly as you would find if you made the measurements with the planets of our Solar System.

However, Freeman was surprised to find that this was not what his results showed. Yes, the objects nearest to a galaxy's centre did whizz around rapidly, but, far from declining with increasing distance from the centre, the rotation curves stayed almost level, right out to the extremities of the disc. The only way such curves could be explained was if there was a lot of invisible material in the outer parts of each galaxy. There it was again—Zwicky's mysterious dark matter.

At the time, this conclusion had a mixed reception in the scientific world. Some astronomers took the dark matter issue seriously, while others thought there was no problem. However, Freeman's eventual vindication was celebrated in October 2012 when, to the delight of all Australian astronomers, he received the nation's highest scientific honour, the Prime Minister's Prize for Science. Like Nobel Laureate Brian Schmidt, Ken Freeman is an astronomer at the Australian National University's Mount Stromlo Observatory. There must be something in the water there.

It was not until 1978, four years after Zwicky's death, that the dust began to settle over the issue. By then, a US astronomer called Vera Rubin had extended Freeman's observations, and realised that each galaxy must be embedded in a giant spherical halo of something invisible that was providing additional gravitational attraction on the stars or gas clouds. Since Rubin's work, the study of dark matter has become a veritable crusade among astronomers. It's hardly surprising, given that we'd like to think we know what the Universe is made of. But this has led to a remarkably detailed view of the characteristics of dark matter—even though its exact nature has remained elusive.

At first, in the wake of these discoveries, there were two competing theories as to what constitutes dark matter. Uncharacteristically for astronomers, the candidates were given rather clever names, being described as either WIMPs or MACHOs. (The former stands for 'weakly interacting massive particles', while the latter means 'massive compact halo objects'. You'd probably guessed that already.) Taking the MACHOs first, the idea was that galaxies might be accompanied by an unexpectedly high number of dark objects like massive planets, dim brown dwarf stars and perhaps even black holes. They would occupy the galaxies' spherical halos, which were already known to be populated by faint stars. Such MACHOs would be very difficult to observe but, if they were present in sufficiently high numbers, might account for the dark matter. The alternative theory was that dark matter exists at the subatomic level and that what we are seeing is the effect of vast swathes of subatomic particles that do have mass but don't interact significantly with the various particles that constitute normal matter. WIMPs, in other words.

It didn't take long to establish that MACHOs couldn't be the primary component. An experiment performed during the 1990s with a 1.25-metre telescope at the Mount Stromlo Observatory found that there weren't enough MACHOs to account for the observed levels of dark matter. The observations were made using a clever technique called 'gravitational microlensing'. You take tens of thousands of repeated images of a single region of sky containing a huge number of distant stars—for example, the central region of our Milky Way Galaxy, or our neighbour galaxy, the Large Magellanic Cloud. If the intervening space were populated by large numbers of MACHOs, you would expect to see a characteristic rise and fall in the brightness of some of the background stars over the period during which the images were made. That would signify that MACHOs had been passing in front of them.

Why would intervening objects brighten the background stars rather than dim them? Because, as we discovered in Chapter 8, the mass of an object distorts the space around it, making it bend. Whereas the bending produced by a nearby object such as the eclipsed Sun simply distorts the apparent positions of stars, more distant objects bend light in the same way that a glass lens would. It's an unusually shaped lens, admittedly, since it's more like the bottom of a wine glass than a normal lens. That means the focusing effect is nowhere near as crisp as in a real lens and results in arc-like images of the distant objects. (Check them out next time you're enjoying a glass of wine—but do make sure the glass is empty first.) Nevertheless, the principle is the same. The distorted space around the MACHO would act like a kind of magnifying glass, enhancing the brightness of any star it passed in front of.

In failing to find an adequate number of MACHOs, the Mount Stromlo experiment put the spotlight firmly on WIMPs as the origin of dark matter, and there it remains today. However, even though we don't yet know what these subatomic particles are, our understanding of their properties has blossomed over the last decade.

The galaxy redshift surveys I mentioned in connection with dark energy also have a major role to play in studies of dark matter, because dark matter exhibits gravity, and we know that the effect of that is to distort the geometry of space. Likewise, visible matter has a distorting effect, but we can allow for this because we can see where it is. This tells us, for example, that the ratio of dark matter to visible matter in the Universe is about five to one, and that dark matter and visible matter actually concentrate together. 'Beacons of light on hills of dark matter' is one eloquent description of clusters of galaxies.

Dark energy and dark matter have quite different properties, and this allows us to disentangle their effects on the geometry of the Universe. For a start, dark energy is a relatively weak repulsive force, while dark matter exhibits a pretty robust level of gravitational attraction. Moreover, dark energy is everywhere—a property of space itself—whereas dark matter occurs in blobs in the vicinity of galaxies. When the large-scale geometry of the Universe is investigated with all this taken into account, we arrive at the 72, 23 and 5 per cent mix of dark energy, dark matter and normal matter mentioned earlier. And it's worth noting that most of that normal matter is actually hydrogen and helium left over from the Big Bang still permeating the Cosmos. The stars in the Universe—all of them—constitute a mere 0.5 per cent of the total, while the stuff from which our world and we ourselves are made—carbon, oxygen, nitrogen, silicon, iron and so on—amounts to a humble 0.03 per cent. If that doesn't put things into perspective, nothing will.

A number of new galaxy surveys are continuing the exploration of space's geometrical distortions in great detail. One, called Galaxy and Mass Assembly, involves observations of 375 000 galaxies with a wide range of different telescopes (including the AAT). Its aim is to produce the most comprehensive database to date of galaxies and their properties as they have evolved over the past one-third of the age of the Universe. Of course, the dark matter halos of galaxies are part of the survey's stock-in-trade.

Knowing that dark matter concentrates wherever visible matter is found suggests another approach to the investigation. Our own Milky Way Galaxy is a giant spiral system of 400 billion stars with associated gas and dust—and dark matter. So dark matter is all around us. How is it distributed? Does it occur in clumps, and, if so, how big are they? And what might we learn from their size? The fact that we are surrounded by stars whose histories reflect the history of our Galaxy as a whole suggests that much can be learned about our Galaxy's past by studying them. As we saw in Chapter 7, this is the basis of the science of Galactic archaeology and has prompted large-scale star surveys such as RAVE and the GALAH experiment. One possible spin-off from these surveys is a more detailed look at how dark matter is structured. Stars move under the influence of gravity, and if we sample large populations of stars the underlying gravity field can be mapped and the dark matter component identified. A likely outcome of this is an estimate of the minimum size of a clump of dark matter. Such estimates have already been made from more limited star velocity surveys, and suggest it may be around 1000 light-years across. This has implications for the temperature—the energy of motion—of the dark matter particles, hinting that they may be warmer than expected, at a few degrees above absolute zero rather than the few tenths of a degree that is usually assumed. As the various surveys evolve, we will see better measurements being made, and it is possible that the temperature of dark matter will eventually be one of the better-determined outcomes.

In the future, this technique of large-scale velocity mapping will be extended to sizeable samples of individual stars in our neighbouring galaxies—stars not yet accessible with existing telescopes. But, with the coming generation of extremely large telescopes mentioned earlier, such observations will be fairly straightforward.

Another powerful way of probing the structure of dark matter in the Universe goes back to the idea of matter distorting the space around it to produce a lens-like effect on background objects. The most dramatic distortions of space come from the most massive objects, and they are not just single stars—or even single galaxies—but clusters of galaxies. Whereas single stars exhibit gravitational microlensing, there's nothing micro about the gravitational effect of a galaxy cluster. Imagine a cluster of galaxies sitting in front of a group of much more distant galaxies in the background. The mass of the foreground cluster distorts the space around it, generating a gravitational lens of the kind we met a couple of pages ago. We've seen that this has the potential to magnify the light of anything in the background, acting as a kind of gigantic natural telescope, so that we can often detect distant galaxies that would otherwise be invisible. But the distant galaxies are also turned into distorted arcs of light by this process, exactly in accordance with the wine glass model described earlier. By analysing these arcs in a statistical manner, it's possible to reconstruct the distortion of space around the foreground galaxy cluster, making a detailed map of the distorted geometry. Then, from the map, the actual distribution of matter in the foreground cluster can be charted accurately.

But here's the clever bit. Gravitational lensing is the result of both visible and dark matter, so the fact that we know where the visible matter is allows us to deduce exactly where the dark matter is hiding. This is an extremely effective technique and confirms that galaxies are embedded in large volumes of dark matter, just as suggested by the other methods. Because of the extreme faintness of the background objects, the work was the province of the Hubble Space Telescope during the 1990s, but today's generation of 8- and 10-metre-class telescopes is capable not only of providing images of the faint background objects but also of obtaining spectra to establish their distances. In the future, it will be extended to very great distances indeed—at which almost all objects are distorted by the lensing effect of intervening matter—using the new extremely large telescopes.

Meanwhile, the Hubble Space Telescope continues to provide new insights into the behaviour of dark matter using the lensing technique. Several recent sets of observations have allowed us to explore the behaviour of dark matter when galaxy clusters collide. While the galaxies and the rarefied gas clouds accompanying them grind to a halt in the pile-up, their respective dark matter halos carry on oblivious to the chaos, a result of the feeble interaction between dark and normal matter. Other experiments have probed the distribution of dark matter over large distance ranges. One compared galaxies with a look-back time (equivalent to distance) of about 3.5 billion years, and others have look-back times almost twice as long. Crucially, differences in the characteristic size of the dark matter cocoons of these galaxies suggest that typical clumps of dark matter have become more fragmented as time has gone on. Such observations involving a wide range of look-back times enhance our understanding of the vital role played by dark matter in the evolution of today's galaxies. The assumption is that dark matter clumped together in the early Universe and its enhanced gravitational field then attracted concentrations of hydrogen from the Big Bang, which eventually collapsed into stars and galaxies. A consequence of this is that dark matter at greater look-back times should be less clumped than it is in the more recent past—which is exactly what has been found.

SCIENCE IN A SPIN

By combining all of these various studies of dark matter, astronomers hope to learn enough about its behaviour for a clear leader to emerge from the various competing models. These have been built primarily by the theoretical physicists who study subatomic particles. But it is in the experimental research accompanying this that the quest for dark matter is taking perhaps its most exciting turn—if you'll pardon the pun—bringing hopes of a real breakthrough. It also leads us straight back to the underlying theme of this book, because those experiments are taking place at one of the most inspiring scientific institutions in the world, and it is an institution that makes visitors welcome.

I'm sure you're aware of the Large Hadron Collider (LHC), the giant atom-smasher near Geneva, which is operated by the European Centre for Nuclear Research. This machine is the successor to the facility visited by our old friend the comedian Ken Campbell earlier in this chapter. That was the Large Electron-Positron Collider, and it occupied a 27-kilometre-long circular tunnel, which was excavated for it in the 1980s. When plans were made for a new machine, the same tunnel was to be used, so there was a lengthy period when physics took second place to engineering on the site. In December 2009, however, the LHC was fired up, and, apart from one early mishap, it has performed as brilliantly as expected.

The LHC's role is to accelerate two streams of subatomic particles such as protons around circular paths in opposite directions, achieving speeds close to the speed of light—and then collide them together. What could be simpler than that? But the technology required to achieve this is little short of astonishing. As one of the largest scientific experiments in the world, the LHC bursts with engineering superlatives. Unfortunately for spectators—or perhaps fortunately—most of the high-energy action takes place deep underground, where the old Large Electron-Positron Collider tunnel now houses twin vacuum tubes containing the particle beams. The plumbing alone is staggering. For example, rope-like skeins of microscopic copper tubing run for kilometres carrying supercooled liquid helium. The vacuum through which the particles travel is ten times lower than the vacuum at the surface of the Moon, meaning that there are fewer residual air molecules for the particles to bump into. And those particles are kept on track by superconducting magnets that are colder than space itself. Some folk have commented unfavourably on the centre's €6 billion price tag for the LHC, but my personal view is that it's amazing they have been able to build it so cheaply.

As you can probably guess, I'm a bit of a fan of the LHC, and to date I've had the opportunity to make two visits there. The first was with the Stargazer II tour, in 2010, when we were hosted by scientists Klaus Bätzner and Quentin King, who graciously fielded all our questions about the machine and its various experiments. A highlight of the tour was lunch in one of the centre's cafeterias, where thousands of enthusiastic scientists and engineers spend their lunchtimes eating, drinking and talking physics. The excitement there was palpable, and it remained with us as we boarded the high-speed train to Paris later that afternoon. We touched almost 300 kilometres per hour on that journey, a minuscule fraction of the almost 300 000 kilometres per second reached by the protons circulating in the collider—but, unlike the protons, we didn't collide with anything.

My second visit was probably a consequence of the first, because by then everyone thought I was an expert on the LHC. (As if . . .) In some ways it was even more exciting, because it sought to tread in Ken Campbell's footsteps by making a down-to-earth TV documentary about this most esoteric of human endeavours. That led our TV crew to parts of the facility that are normally out of bounds to visitors, such as the inside of some of the experimental control rooms. And, once again, it led us to the cafeteria, where the buzz was captured on camera with the help of a handful of Australian and Kiwi scientists who were happy to give us a lunchtime taste of their work.

In an echo of the high-speed train ride, that trip also led us to a ridge in the Jura Mountains behind Geneva, to a place called Col de la Faucille, which boasts a gut-wrenching rollercoaster ride intended to simulate an Olympic luge track. The idea was to give viewers a hint of what it might feel like to be a subatomic particle circulating in the LHC, by mounting a camera on one of the cars while we presenters took turns expounding the details of impossible sideways accelerations around the ride's hairpin bends. This time, we reached speeds only of around 40 kilometres per hour—but, when your bottom is just a few centimetres above the track, I can tell you it feels a lot like the speed of light.

All the marvellous engineering notwithstanding, it is the LHC's potential for scientific discovery that excites physicists and astronomers. By smashing together particles such as protons, the LHC is effectively acting as a gigantic super-microscope, probing matter on the smallest scale by examining the cascade of subatomic debris released in the collisions. It is exactly analogous to breaking things into successively smaller pieces until they can't be broken any more. Then you know you are dealing with the fundamental building blocks of matter. The collider's first job was to test something called the 'standard model', which is a complex hierarchy of those fundamental building blocks. It consists of twelve indivisible particles of matter that range from the familiar (like electrons) to the decidedly peculiar (like top quarks).

Then there are four force particles carrying three of the four fundamental forces of nature. Particles carrying forces? Welcome to the strange world of subatomic physics. And why four particles to carry three forces? Because one of them, the weak nuclear force, is greedy, and needs two. And what happened to the other fundamental force? That is gravity, and it is actually beyond this standard model at present, because we don't yet have a satisfactory theory of quantum gravity describing the way it acts on very small scales.

The standard model does, however, suggest the existence of a particle that had not yet been found at the time of my visits—the Higgs boson, a 1960s postulate of physicist Peter Ware Higgs and his colleagues at Edinburgh University. Boson is the generic name given to the force carriers mentioned above, and the Higgs boson is thought to endow all of the other standard model particles with the property of mass, so it's fundamental to the model. And one of the first tasks of the LHC was to discover where it could be hiding. Not in terms of its position in space but where in the range of energies that characterise the size of things in the subatomic world—in a similar manner to the way spectrum lines characterise chemical elements that are emitting light. By the end of 2011, the first hint of the Higgs had been announced. The evidence was spotted not just in one but in two of the ongoing experiments being carried out at the LHC, an essential requirement for the finding to be considered valid. But confirmation of the Higgs' existence and a measurement of its mass took much more work, and it wasn't until July 2012 that a formal announcement was made. Even then, the statement was couched in the most cautious of terms, as meets a discovery in a science that depends so heavily on probabilities. The likelihood is, though, that the Higgs boson has now been found, and physicists the world over can get on with more detailed investigations of the standard model with renewed confidence that it correctly describes what nature is up to.

It is another area of physics—a postulated extension to the standard model called 'supersymmetry'—that carries the hopes of astronomers regarding the nature of dark matter. The idea behind supersymmetry is that each particle in the standard model has a massive 'shadow particle' that has not yet been detected. So there might be an entire suite of undiscovered particles that together constitute a supersymmetric version of the standard model. The expected characteristics of at least one of these shadow particles would exactly fit the bill for dark matter. It would be massive but would not interact with normal matter, except through gravity. Finding supersymmetry at the LHC is turning out to be just as difficult as the hunt for the Higgs, with most of the simpler models having already been ruled out. But astronomers and physicists remain hopeful of a breakthrough in this area. And the odds are that when the first announcement of the true nature of dark matter is made, it will come not from a telescope but from a particle collider.

Meanwhile, a recent incident at the LHC provides a tantalising glimpse of the reception that awaits the discoverers of any hint of new physics. In September 2011, the extraordinary announcement was made that exotic subatomic particles called 'neutrinos' had been clocked travelling ever so slightly faster than the speed of light. Although it's an idea beloved of science-fiction writers, most folk know that faster-than-light travel is firmly prohibited by Einstein's Special Theory of Relativity, because accelerating an object to the speed of light requires infinite energy. This tried and tested rule is at the heart of our understanding of the Universe, but it is just possible that the physical world actually consists of more than the three dimensions of space and one of time that we perceive—and might have hidden higher dimensions. If that is the case, then objects like neutrinos may be able to take short cuts through these higher dimensions, arriving fractionally before they otherwise would, while still obeying the laws of special relativity. Thus, the result of the LHC experiment was viewed as a tantalising hint of the existence of such new physics and attracted intense media interest.

The overwhelming view of the world's scientists, however, was that the faster-than-light effect would be shown to have an explanation entirely within the realm of normal physics. And, little more than six months after the original announcement, that's exactly what happened. A tiny measurement error apparently caused by a loose connection had resulted in an incorrect calculation of the neutrinos' flight times. Red-faced, the head of the experiment resigned his position, despite having been ultra-cautious when he'd first presented his astonishing result.

So, RIP new physics? Not necessarily. It remains possible that, some day, we may find similar evidence that does stand up to scrutiny, with implications for science that are truly staggering. The discovery of hidden dimensions would allow new thinking on a wide range of topics, including the origin of that other big cosmic mystery, dark energy. And, as we shall see in the final chapter, such a discovery could feed directly into our understanding of reality at the most profound level.

For those of us cheering them on from the sidelines, it is a most exciting time to be watching the work of scientists and engineers at the LHC. There is unprecedented interest among ordinary people in what is happening there, and, as a result, the world of physics may even be starting to lose its image as the exclusive province of nerds. Books, articles and even TV segments featuring insane rollercoaster rides are all helping to satisfy the public's hunger for information. And here's a suggestion for one more educational item in this field. You can probably guess what it is. Forget the pop-up Big Bang. What we need now is a pop-up book of new physics, complete with faster-than-light particles. With all those hidden dimensions, the folded-paper work should be easy.

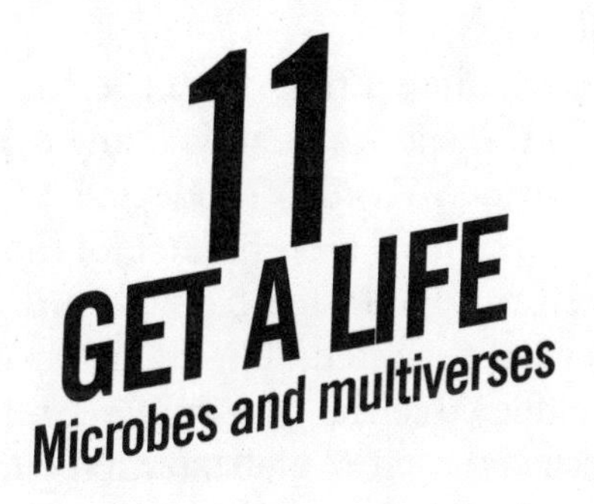

11 GET A LIFE

Microbes and multiverses

Of all the great astronomy travellers of the twentieth century, there is one who stands out not only for his science and his air miles but also for his incisive wit. Richard van der Riet Woolley spent much of his life journeying between the United Kingdom and South Africa, but he also lived for significant periods in the United States and Australia. He is best known as the eleventh astronomer royal, a post he held from 1956 to 1971 with great distinction—and for which he received a knighthood in 1963.

Woolley was also known for his classic one-liners. Most famous is the comment he made on his arrival in the United Kingdom to take up his new job, early in 1956. Asked by a journalist what he thought of the prospects for space travel, the astronomer royal replied that talk of space travel was 'utter bilge'. More than half a century later, that acerbic put-down still echoes through the folklore of British astronomy—even though it was most likely the media hype surrounding space travel that he was criticising.

While 'utter bilge' might have been an appropriate (if undignified) response at almost any earlier time in history, what called Woolley's judgement into question was the launch, only 21 months later, of the world's first artificial satellite. Of course, that was hardly something he could have anticipated. The Soviet space program had been conducted in the utmost secrecy, and *Sputnik*'s debut astonished the entire globe. But Woolley must have been aware of the competing US *Vanguard* program, and no doubt shared the embarrassed confusion of its scientists and engineers in the wake of the Russian triumph.

Before Woolley became astronomer royal, he was for seventeen years director of the Mount Stromlo Observatory—then still known as the Commonwealth Observatory. That venerable Australian institution made worldwide headlines in our own era, when all its heritage buildings—including six historic telescopes—were destroyed in the savage bushfire of 18 January 2003. But in Woolley's time, too, fire came close to consuming the observatory. On 5 February 1952, in the middle of a particularly hot summer, lightning started a fire that eventually reached the neighbouring pine forest and swept through it to the observatory's workshop, which was completely destroyed. As in the 2003 inferno, the fire front moved with devastating speed, and it was left to the staff to fight it with whatever resources they could lay their hands on. That Woolley was not particularly impressed with the outside help they received was evident from a comment he made some years later:

> When the fire had passed [through], up came a barrel of beer, and sandwiches, with the compliments of the Hotel Canberra. Finally there came the Fire Brigade, who drank the beer. Then the Minister for the Interior, who started giving the fire-fighters instructions.

Point taken.

Modern-day historians Tom Frame and the late Don Faulkner of Mount Stromlo have uncovered further examples of the way Woolley used words to devastating effect in confronting adversaries—but in a manner you can't help liking. They have revealed that the contemptuous and, with hindsight, often grossly misjudged one-liner was part of his stock-in-trade. In 1947, for example, at a high-profile conference in Perth, Western Australia, he was asked where he thought the science of radio astronomy would be in ten years' time. Australia was already emerging as a major force in this new field, so Woolley's radio astronomy colleagues might have been forgiven for anticipating a few words of praise. They were decidedly unimpressed, however, with his answer: 'Forgotten.' The fall-out from that misguided remark took years to settle. And a reliable eyewitness account of a conversation held five years later between Woolley and the great French-US astronomer Gerard Henri de Vaucouleurs (who worked at Mount Stromlo in the 1950s) revealed a similarly withering response. De Vaucouleurs had just completed his seminal work on the local supercluster of galaxies and excitedly asked Woolley where he thought it should be published. The *Astronomical Journal*, perhaps? '*Punch*,' replied Woolley.

I shudder to think what Woolley would have made of some of the science I'd like to tell you about in this chapter. It's hardly likely that someone who refused to see (or pretended not to see) the value of developments such as radio astronomy would have embraced the comprehensive and sometimes speculative view of the Universe we have today. The science of the stars has become a holistic enterprise, a broad-brush undertaking that goes to the heart of our perception of ourselves and our place in the Cosmos. As we saw in the last chapter, it also goes hand in glove with other sciences, such as particle physics. All of that contrasts strongly with the situation even as recently as 1976, when Woolley retired from his job as director of the South African Astronomical Observatory.

In exploring some of the contemporary astronomical issues that might have been 'utter bilge' to Woolley, we can do no better than start with perhaps the most basic question we can ask: 'Where did we come from?' Today, we have a pretty good idea of the answer, even if, at the outset, it does sound uncannily familiar. However, as creation stories go, this one has much more to recommend it than most.

THE FACTS OF LIFE

In the beginning, there was . . . nothing. Then it exploded, and space and time were created in a searing flash of light, together with a lot of dark matter and superheated gas. The gas was mostly hydrogen and helium, while the dark matter was, well, dark matter. Not a terribly auspicious start for an infant Universe aspiring to a rich and varied structure. But then, gravity gently took control, bringing into being stars, galaxies and planets, and eventually continents, animals, human beings and *The Simpsons*. Not to mention black holes, quasars and supernovae.

It's extraordinary—and perhaps not a little absurd—that we have such a complete and detailed picture of the history of the Universe since its birth 13.7 billion years ago. Admittedly, we don't understand the mechanism that triggered the Big Bang itself, if indeed there was a mechanism. One bizarre suggestion is that it resulted from a runaway intensification of quantum fluctuations in nothing. Make of that what you will. Or if, as relativity suggests when the theory is applied to the Universe as a whole, time started with the Big Bang, there's actually no need for cause and effect. The Big Bang just happened.

We do know what happened next, though, and that is partly because, as I have noted before, astronomers are gifted with the ability to see directly into the past by virtue of the finite speed of light. For a few hundred million years after the flash of the Big Bang subsided, there was darkness. But then, the first generation of stars lit up the sky. While we haven't yet observed them, we have a pretty good idea what those stars looked like. Short-lived, and with a prodigious energy output from the nuclear reactions taking place within them, these monsters had perhaps 100 times the mass of the Sun. Deep inside their cores, hydrogen burned and forged new elements such as carbon, nitrogen, oxygen, silicon, calcium and iron.

After only a few million years, these first stars ended their brief lives in monumental explosions, achieving temperatures high enough to make much rarer elements—gold, silver, uranium and others. In that way, they endowed their surroundings with all the raw materials of our world (and a few precious metals), and over time those elements found their way into new generations of stars. The process was repeated in a continuous cycle of stellar life and death, resulting in a gradual increase in atomic elements throughout their environment. In the geeky jargon of Chapter 7, the metallicity of the interstellar medium was enriched.

In the cold of space, some of those elements began reacting chemically with one another to create molecules such as water (in the form of ice crystals) and silicates, which condensed into solid grains of dust. And so the gravitational collapse of one particular cloud of gas, ice and dust 4.6 billion years ago led to the formation of a cluster of stars, one of which was our Sun. With a flattened disc of dust and gas rotating around the newborn Sun, the stage was set for a rapid transformation. Fluffy balls of dusty material steadily coalesced with others under their mutual gravity to build solid lumps of rock and eventually infant planets by the process called 'accretion'. Within a few million years, the Sun and its dust cloud had become the recognisable forerunner of our Solar System—a shining star with a family of orbiting planets, fringed by left-over icy debris in the form of comets.

The details of how we wound up with four rocky planets close to the Sun (Mercury, Venus, Earth and Mars) and four gas giant planets further out (Jupiter, Saturn, Uranus and Neptune) are broadly understood today. The best models also suggest there may have been a fifth gas giant, between Saturn and Uranus, that was ejected within the first 100 million years or so, perhaps to become one of those orphan planets, or FFLOPs, we met in Chapter 2.

Notwithstanding the sad story of that wayward child, interesting things started happening on the third planet. Bombardment by icy bodies from the outer Solar System had given it a vast reservoir of water, kept liquid by the planet's proximity to the Sun. As you may know, the Earth orbits in the Sun's Goldilocks Zone, where it's not too hot and not too cold but *just right* for liquid water to exist. And some rather suggestive molecules had also found their way onto the planet from the depths of space, locked up in meteorites. They were amino acids—the building blocks of terrestrial life.

With water to act as a solvent bath and complex carbon-based molecules in abundance, it may have been inevitable that life would begin on the Earth. We don't yet know how that happened; indeed, at the most basic level, we find it hard to define exactly what life is. Most definitions cite self-sustaining, self-replicating organisms that are capable of Darwinian evolution, but others are much broader. One of them even suggests that life's main characteristic is simply that it modifies its environment. Not always for the better.

What we do know is that the first microbes emerged within little more than a billion years of our planet's creation. Another billion years or so found a planet with an oxygen-rich atmosphere, made that way by the presence of living organisms. And, after a further 2.4 billion years of evolution, those rudimentary life forms have turned into the seven billion specimens of *Homo sapiens* that rule the planet today. And, of course, Homer Simpson.

So much for our origins. It rather sets the record straight regarding our importance in the grand scheme of things. Despite all the trappings of 21st-century life, we're still just a bunch of evolved microbes, roaming over the surface of a cinder left over from the Sun's formation.

Our ultimate fate as a species is no less humble. Ignoring the prospect of a Big Rip, which is so far down the track as to be irrelevant, there are a number of possible endings for humankind. Assuming we don't wipe ourselves out by climate change, overpopulation or war, our most likely demise will be at the hands of a runaway virus, a supervolcano, a rogue asteroid or a nearby supernova explosion. At least some of these scenarios have a technological solution, assuming we can act quickly enough. But what we will not survive is the Sun's decline into old age as a red giant star, five billion years hence. Then, the Earth will be swallowed up by our star's expanding outer envelope, and it will be curtains for any life on the planet. That is so far in the future, though, that the evolutionary processes that have brought us to our present state will have taken us much, much further. Who knows? By then, our distant descendants may have colonised the Galaxy and will barely notice the demise of one tired old star.

The evolution of the Universe and its contents is a grand and marvellous story. Stars are born and die, and the material from which they and their planets are made is recycled. And we are an integral part of that process. Something like one-third of the atoms in your body were created inside a long-dead star. And those same atoms, long, long after your death, will become part of the expanding cloud of gas that will follow the Sun's demise. They may find their way into the planets of new stars and perhaps, in a few billion years, into new living organisms of unimaginable kinds. In that sense, parts of you will live on, probably forever.

Talking of recycled atoms, I do take some slight comfort in the fact that in the much shorter term—say, over the next 50 000 years—atomic particles in my earthly remains (whatever form they take) will still be busy doing things. In particular, the decay of my body's carbon atoms from an unstable type called 'carbon-14' to a stable type called 'carbon-12' will cheerfully keep going for thousands of years after my death. That, of course, is the principle of carbon-14 dating for organic material, mentioned back in Chapter 3: you look at the relative proportions of the two types in order to determine how long the nuclear decay has been going on. But it also means that bits of all of us will be alive and kicking in the atomic world for an extremely long time.

There are some aspects of human life that are likely to change on a much shorter timescale. We live in a world where technology is exploding with exciting new developments. Many of these relate to medicine and impact directly on our day-to-day lives (think functional MRI scans, laser surgery, laparoscopies and the like). But an even bigger revolution is just around the corner. We now understand many of the processes that are involved in ageing—the shortening of chromosome telomeres, for example—and it is only a matter of time before these are addressed with tools such as nanotechnology. We already live in an era in which the expected lifespan in the western world increases by about three years for every decade that passes. But it's possible that bioengineering might eventually be able to extend life almost indefinitely, so that death is a rare occurrence and perhaps even a matter of choice. That would clearly require the introduction of a whole new suite of ethics, not least to work out how the world's resources should be shared among its burgeoning population.

An even more outlandish situation could be looming on the horizon. With the extraordinary advances in computing seen over the past few years, we are not too far from machines whose capabilities can rival the human brain. Some researchers predict that we will be outsmarted by machines well before 2050. Such artificial intelligence could be used to supplement our mental capabilities, perhaps with a plug-in device similar to a memory stick. No, I'm not making this up—bionic devices are already paving the way. But there is a further development of that, envisaged by informed future-watchers like the science-fiction writer Jack Dann. Notwithstanding the incredible complexity of the brain, Dann thinks that someday down the track, 'We might exist in a silicon world after we're dead; our personalities, thoughts, memories, and desires might be as portable as computer disks, and our bodies disposable as paper towels.' Wow. Paper towels. That would certainly give a new slant to your body image. It also puts the spotlight right back onto the marvellous organ that these supercomputers are supposed to replace.

BRAINS AND BIOMARKERS

The human brain is a truly extraordinary machine. With its 100 billion neurons (about the same number as there are stars in an average galaxy) but with a vastly greater number of interconnections between them, it is the most complex entity known. It reached its current state of evolution in Africa some 200 000 years ago, and it's still evolving. Curiously, in relation to body size the average brain has shrunk slightly over the past 15 000 years. Better diet, more efficient brain processes and the effect of complex social structures in enhancing the survival prospects of the least fit have all been proposed as reasons for this. The point, though, is that it is this amazing organ that gives us our sense of self and a consciousness of the world around us. Whether that consciousness would be preserved in the kind of artificial intelligence machine proposed by Jack Dann is still a matter for debate among neuroscientists and philosophers. But it is consciousness that allows us to ponder our place in the wider scheme of things.

Most of what our distant ancestors knew about their environment was the product of a grand illusion—that we live in a world that is essentially flat and absolutely stationary. A number of things combine to promote this overwhelming illusion of *terra firma*. The most obvious is that the Earth is very large compared with our own physical dimensions, but a fellow conspirator is the insistent pull of gravity. This overpoweringly defines for us the direction to the centre of the Earth—which we perceive simply as 'downwards'—rather than some more-significant direction in space, such as the line along which the planet is travelling. And the Earth's atmosphere tricks us into believing we live in a benign cosmic environment that stretches above us *ad infinitum*. In reality, the atmosphere is that thin membrane separating us from the black harshness of space, which future space tourists will marvel at. It provides us with the 78, 21 and 1 per cent mixture of nitrogen, oxygen and argon that we need to breathe, but we only have to climb to a height of 10 kilometres to be above 75 per cent of its mass. There's no wonder our forebears believed we were at the centre of everything, with the stars and planets appearing to circulate around the Earth.

But those ancient brains eventually got to grips with the illusion, and since the time of Copernicus and Galileo we have known that we're not actually at the centre of the Solar System. As astronomy has progressed, we have come to realise that we're not in any kind of special position, whether it be within our Milky Way Galaxy, the Local Group of galaxies or the wider Universe. This notion of being nowhere special is called the Copernican principle, and it underpins modern cosmology. One of its consequences is the notion that the everyday environment that we regard as commonplace is probably not unique, though it is undoubtedly rare in the Universe. That has led to the emergence of the new field of astrobiology, which brings together various scientific disciplines such as biology, astronomy, geophysics, planetary science, chemistry and meteorology to investigate whether living organisms might exist beyond the Earth. At present, we simply don't know.

Astrobiology is currently riding a wave of popularity in the world of science. I suspect biologists like it because it allows them to talk professionally about the Universe, while astronomers like it because it allows them to talk professionally about sex. More seriously, it has brought the question 'Are we alone?' into much sharper focus. While most astrobiologists are optimistic about the prospects of finding life elsewhere (encouraged, perhaps, by the discovery of living organisms in virtually every nook and cranny of the terrestrial environment), the possibility still remains that the Sun could be unique among the 10^{22} stars in the observable Universe in harbouring life among its planets. That would be both extraordinary and humbling—but also impossible to verify.

In terms of the search for life, astrobiology is progressing on two main fronts. Within the Solar System, robotic spacecraft are making ever-deeper forays into the environments of our neighbouring planets. For example, as we saw in Chapter 9, we now believe Mars' northern hemisphere was once covered by a shallow ocean. With water such an effective medium for nurturing and sustaining life, there seems every likelihood that rudimentary organisms once flourished there. Perhaps someday quite soon we will find Martian marine fossils from that era, almost four billion years ago, giving us a whole new realm of biodiversity. That could even lead to a revolution in pharmacology, with as-yet-unknown molecules from Martian organisms providing effective therapies for Earthly diseases. Astropharmacy—it has a nice ring to it, don't you think? Remember where you heard it first.

More intriguing are recent measurements suggesting that some of the rocky moons of Jupiter and Saturn have outer surfaces that are merely thick crusts of ice floating on global oceans of liquid water. The water is kept liquid by internal heat generated in the gravitational squeezing and stretching of the moons by their parent planets—for this is well beyond the Goldilocks Zone. Strong supporting evidence has come from the *Cassini* spacecraft orbiting Saturn. It has analysed ice geysers erupting from Saturn's moon Enceladus and found minerals suggesting they originated in a large body of water with a similar level of salinity to the Earth's oceans. Who knows what might be thriving in the cold darkness beneath?

Also intriguing are the hydrocarbon seas of Saturn's giant moon, Titan. This strange world has an icy surface blanketed with a thick foggy atmosphere and clouds of methane and ethane. On the ground there is wind and rain, but at an average temperature of –180 degrees Celsius the raindrops are not water but liquid methane—as are the seas and lakes spotted by *Cassini* in Titan's polar regions. The chemistry of Titan's atmosphere has recently revealed an enigmatic hint of exotic microbial activity, a tantalising glimpse that will be confirmed only with future space missions.

Finding primitive microbial life on Titan would be doubly significant, as the utilisation of methane rather than water as a solvent would be highly suggestive of a second genesis, in which life has started independently of terrestrial life. (That would not necessarily be true of water-based life forms that might be found on Mars.) A second genesis would provide a clue that life can start whenever the right conditions are found—something we don't know at the moment. That would, in turn, suggest that life is common throughout the Universe.

Equally significant would be signs of life revealed by telescopic investigation of distant solar systems, which is the other main exploration front of astrobiology. For this, we have to rely on the evidence of spectroscopy—the analysis of light coming from remote planets. Of the more than 800 extrasolar planets currently known, only a handful have so far been investigated for the presence of biomarkers—tracers of life—in their atmospheres, but that will change dramatically with the advent of extremely large telescopes in the next decade or so. Potential biomarkers include the simultaneous presence of oxygen and methane in relatively large amounts, a situation that is unlikely to occur without life processes—and there are several more.

Still more spectacular would be the discovery of what might be called 'technomarkers'. For example, certain pollutants found in the atmosphere of an extrasolar planet could signify large-scale industrial processes rather than natural chemistry. That would reveal the presence of thinking beings tinkering with their environment—just as we did with the release of chlorofluorocarbons into our own atmosphere. And perhaps one of the claims of my radio astronomer friends spruiking their proposed Square Kilometre Array is more than just a joke. This southern-hemisphere radio telescope will be the most powerful in the world when it becomes fully operational in the 2020s, and I'm reliably informed that it will be able to detect airport radar signals at a distance of 50 light-years. That could be the ultimate technomarker.

With all such possibilities for investigating extraterrestrial intelligence, however, there is one overriding caution: you have to be looking for your hypothetical intelligent beings at *exactly* the right time. The length of time that the Earth has been occupied by a technological civilisation—a few millennia at most—is but one-millionth of the age of the planet. And, compared even with that, the era of industrialisation is fleetingly brief. Moreover, who knows how long we might last? If technological civilisations are short-lived phenomena, the odds against finding them are greatly increased. I have to say, though, that I don't think this should stop us looking.

THE MEANING OF LIFE . . .

With millions of candidate solar systems within the Sun's neighbourhood, the search for life is likely to be one of the great ventures of 21st-century astronomy. It is interesting to consider how that search will address the final question I'd like to deal with in this chapter: 'Why are we here?' I'm not suggesting that if we do find aliens we should ask them, but that we will then know with certainty that we are just a small part of something very large indeed. While we are still at the top of the tree in terms of known life forms, it is likely that the Copernican principle will apply here—that we are nothing special among the intergalactic throng. In some ways, perhaps, humankind's highest purpose is to seek to understand our environment as completely as we can. That includes investigating all the mechanisms of the Universe, most of whose secrets are still hidden. But then, of course, I would say that, being an astronomer.

There is one further curious aspect of this that begs discussion. Many scientists and philosophers have noted that the Universe appears somehow finely tuned for living organisms to exist. That is to say, if the fundamental constants of nature were different from those that we observe, we wouldn't be here to observe them. Tweak them slightly one way and stars never form; tweak them the other and the Universe collapses into itself before anything useful has had time to happen. In 1999, the United Kingdom's astronomer royal, Sir Martin John Rees, highlighted the narrow limits within which these various quantities must fall, in a book provocatively entitled *Just Six Numbers.*

So how could this fine tuning have occurred? It's no accident that Rees is also a leading proponent of the notion of the 'multiverse'. The idea of multiple universes is the ultimate extension of the Copernican principle and, at first sight, seems to make no sense. We normally define the Universe as being everything we can detect or know about. It includes all of space-time—whether it's visible to us or not—and all fundamental particles and forces. How can we possibly talk about *other* universes? The answer to that lies at the very frontiers of knowledge. A few pages ago, in connection with the debunked claim of faster-than-light neutrinos, I noted that reality might consist of more than just the three dimensions of space and one of time that we perceive around us. String theory, for example, postulates that vibrating strings—which to us look like subatomic matter and force particles—actually occupy more than just the four dimensions we can see. It requires the existence of several dimensions (the number varies from one version of the theory to another), and the hidden dimensions are assumed to be compactified, or rolled up, in such a way that we can't normally detect them. They may, however, be revealed in collisions of subatomic particles at the highest energies, and that's why physicists are looking to facilities like the LHC to provide the first evidence of their existence—if, indeed, they do exist.

You may also have come across a relative of string theory called M-theory, or brane theory, which suggests that there are larger hidden dimensions in which our normal four-dimensional space-time is immersed. The M has nothing to do with James Bond and probably stands for 'membrane' (hence the abbreviation to 'brane'), although, to be honest, no one seems to know for certain. However, it does convey the right sentiment. Imagine holding up a single page of this book in front of you—as you may well be doing right now. The page is a two-dimensional object located in the three-dimensional space around it. In a similar way, M-theory envisages our Universe residing on a three-dimensional brane that sits in a higher dimensional space known as the 'bulk' (unless you're Doctor Who, in which case you'd call it the 'void'). In such a theory, it's possible to imagine the bulk being populated by other branes containing, yes, other universes—a bit like the other pages in the book, only they're not joined at the spine, and there are gazillions of them. They could even collide with one another from time to time, producing a big bang event in one or both of the unfortunate branes. (I'll leave you to work out how to simulate that yourself, mentioning only that this is officially called the 'ekpyrotic model' of the Big Bang.) For my money, the idea of colliding branes kicking off the Universe has much more to recommend it than runaway quantum fluctuations. Everyone can understand a shunt.

One way that we might gain at least a hint about the existence of other universes is if they have left their imprint in the CMBR, the echo of our own Universe's Big Bang. With a new map of the CMBR promised soon from the European Space Agency's *Planck* spacecraft, this will be an exciting field of investigation.

The point at issue here, though, is that these other universes might have quite different physical constants from each other, and not necessarily ones that would allow life to evolve. But our Universe *has* evolved life—at least once. This is at the heart of something called the 'anthropic principle', which, like string theory, comes in a variety of flavours. The main point is that we are here because the Universe is like it is. Or, conversely, the Universe is like it is because we are here to observe it. So, the anthropic principle provides perhaps the only cogent response to the question 'Why are we here?' The answer: 'We're here because we can be.' To be honest, though, I think the eloquent José put it far more profoundly back in Chapter 10. 'Things have been much better since the Universe was here.'

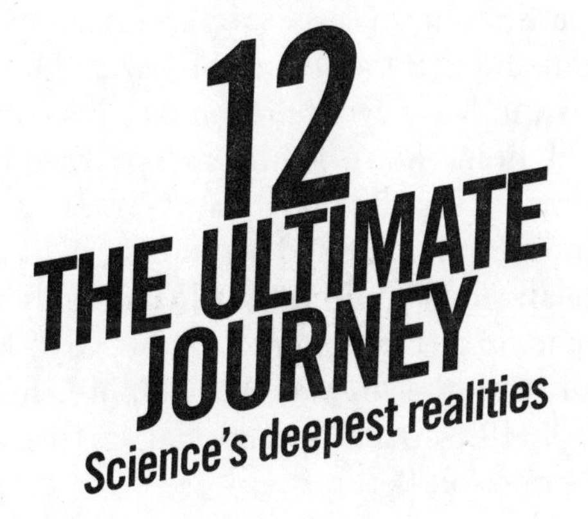

12 THE ULTIMATE JOURNEY

Science's deepest realities

The noise in the cockpit is deafening. Even throttled back for the cruise, the Lancaster's four engines fill the aircraft with sound. Everything rattles. The thin air outside has become a roaring cascade around the fuselage, as whirling propellers pull it ever onwards through the night air. Onwards—of course—to Berlin, where its lethal cargo will be released, in the hope of tipping a finely balanced conflict towards freedom and democracy.

Inside, the air is spiced with an acrid blend of warm electronics and hot oil. Not that the crew can smell it. At their height of nearly 6 kilometres they would asphyxiate without the oxygen masks that cover their faces. By the dim glow of instrument panel lights and starlight glinting through the Perspex cockpit canopy, the five crew members in the forward cockpit area catch occasional glimpses of each other. Out of sight, further down the aircraft, are the mid-upper and tail gunners, isolated and ever vigilant at their posts—and struggling, as always, to keep warm. All communication is via the aircraft intercom.

In command is Pilot Officer Frederick Horace Garnett, known to his crew as Fred and to his two adoring sisters back home in Yorkshire as 'our Freddy'. A few months short of his 22nd birthday, Garnett is a veteran with eight operational missions to his credit. And he's flying a relatively new aircraft, delivered only six weeks ago. In that time, Lancaster ZN-J of No. 106 Squadron, based at Metheringham Airfield, in Lincolnshire, has flown half-a-dozen operational sorties, all—like tonight's—raids on Berlin.

Garnett's crew is a mixed bunch, but they're all good chaps. Among them is one of the youngest serving members of Bomber Command, Sergeant John Withington. He's only eighteen years old, but he knows his job as an air gunner. Immediately behind Garnett sits his navigator, Flight Sergeant Theo Thomas, hard at work throughout the flight, his illuminated plotting table shielded by a black-out curtain. Air navigation has progressed by leaps and bounds in recent months, with a new-fangled radio device, code-named H_2S, that can see through clouds beneath the aircraft to give an impression of the underlying terrain on a cathode-ray tube. Land or water, built-up area or open countryside—H_2S can reveal the difference. (Rumour has it that the code name originated in a stuffy, high-ranking officer's comment that 'it smells like something bad' when he was first told of the invention.) But when all else fails, the tried and tested method of finding your way is still to navigate by the stars. With his bubble sextant, astrocompass and star position tables, Flight Sergeant Thomas has an affinity with the night sky. Truly, Garnett is travelling with an astronomer.

As with many of his colleagues, serving with Bomber Command wasn't Garnett's first preference when he joined the Royal Air Force Volunteer Reserve. The idea of inflicting horror and devastation on a civilian population went against all his values, and still does, but there had really been no choice in the matter. And now, with Churchill sanctioning an all-out onslaught on Berlin in an attempt to crush the heart of the Nazi regime, all available aircraft and crew are being mobilised.

But tonight, there's a new spirit of optimism. When Garnett and his crew took to the air from Metheringham, shortly after midnight, the new year was barely 24 hours old. There's good reason to hope that this will be the last year of the war. Four years of hardship is enough. The whole crew relishes the opportunity to play a part in hammering one more nail into the coffin of a delinquent regime. And they're not alone. Tonight, ZN-J shares the flight path to Berlin with no fewer than 420 other Lancasters. Tonight, Hitler's 1000-year empire will be made just a little shorter.

Garnett and his crew are deep in enemy territory before there's the first hint of trouble. Last night there were no Allied bombing raids into Germany, only coastal mine laying—so the Luftwaffe night fighter crews are well rested and ready for the fray. As the bomber stream approaches Hanover, shadowy forms begin weaving their way into it. H_2S is switched off, because there's a suspicion that the German night fighters now have the ability to home in on the radio signal it emits. And then, over the River Weser, things suddenly hot up. Out of nowhere, an explosion rips through a fellow No. 106 Squadron aircraft, and down it goes. Hard to tell whose kite it is, poor buggers.

But now, sickeningly, it's ZN-J's turn. Again, from nowhere, with not a single fighter aircraft within sight of the Lancaster's crew, cannon shells pierce the fuselage, slicing control lines, hydraulic pipes, fuel lines—and blood vessels. Robbed of control, the stricken aircraft starts to spin. Alex Elsworthy, the tail gunner, seems to have got out—is that a parachute billowing below? Someone else is out—John Withington in the mid-upper turret? Inside the Lancaster, they are twisting, twisting, twisting. Unable to move because of the centrifugal acceleration of the spinning aircraft, the five remaining crew are pinned to their seats. And now, they are falling from the sky . . .

BEREAVEMENT

The weapon that brought down ZN-J, at Otternhagen, near Neustadt am Rübenberge, in the small hours of 2 January 1944, was almost certainly *Schräge Musik*, an oblique, upward-firing cannon fitted to German night fighters to allow them to strike the Lancasters from below. It placed the attacking aircraft beyond the reach of the Lancaster's three defensive gun turrets and was devastatingly effective in the Battle of Berlin between November 1943 and March 1944. During that period, over 1000 Allied aircraft were lost, more than twice the number predicted by Air Marshal Sir Arthur Travers Harris when he had sold the idea to Churchill of a sustained bombing campaign on Berlin. The carnage was one reason why the war in Europe dragged on until May 1945.

Lancaster ZN-J was just one of 28 Allied aircraft that were shot down that night, almost all by night fighters. Most had a similar, sad outcome. Garnett and five of his crew lost their lives. Withington, one of the youngest RAF casualties of the war, is believed to have died as a result of a failed parachute, after he had tried to help other crew members with theirs. Only one man survived—Sergeant Alexander E. Elsworthy, the tail gunner, who was subsequently interned as a prisoner of war.

Back in the United Kingdom, a sombre telegram arrived at the Garnett family home, in Lidget Green, Bradford, with the news that Pilot Officer Frederick Horace Garnett had been posted as missing in action. The effect, quite simply, was devastating. Fred was the only son of Arthur Garnett, a baker, who had himself seen service in the First World War. Fred's mother, Margaret, had succumbed to tuberculosis when he was only five years old, so he had been brought up mainly by his two older sisters, Marjory and Dorothy. The loss had made them a close-knit family, and their pride had known no bounds when Fred was selected for pilot training, in 1941.

Like so many young RAF personnel, Fred was posted to Canada for his flying tuition. That kind of travelling was unheard of, certainly among this branch of the Garnetts. Although Arthur and his kin would have been unaware of it, modern-day genealogy reveals that the family had lived within a couple of dozen kilometres of Lidget Green for at least 350 years. They seldom went anywhere. Arthur's sojourn overseas during 't'First Lot' (as Yorkshiremen of his generation disparagingly labelled the First World War) was very much the exception to the rule. When Fred had become an officer, in 1943, it was as if he had graduated from university with a first-class honours degree—something else that had never happened to a Garnett. His future prospects had seemed limitless.

In many respects, the Garnetts were a typical wartime family—and typical of families on both sides of the conflict. Marjory and Dorothy had clerical jobs but contributed to the war effort as members of the Auxiliary Fire Service. Marjory's husband, Bert, was in a reserved occupation—a post office telephone engineer—but he also served in the Home Guard. Arthur, in his mid-fifties, was excused service in 't' Second Lot', but Dorothy's fiancé, Olaf, had served in North Africa with Montgomery's Desert Rats, and was currently in Italy.

In the short term, at least, there was really only one way to lighten the profound grief that had overwhelmed them all following Fred's death. So Marjory and Bert made a decision and then got on with it. The result of that was me.

THAT OLD-TIME RELIGION

You may well wonder why I have written about these events in a book that is supposed to be about the science of the stars. Partly, it is because I wanted to pay homage to my forebears in a way that is rather unfashionable in our society these days, although it remains a firm pillar of many others. But I'm also interested in a question that seems increasingly relevant today: whether science, in all its extraordinary manifestations, has anything to say to the bereaved. Of course, bereavement can encompass many different kinds of loss, but I'm thinking here specifically of the death of a loved one. And, while it's clear that any given individual's response to bereavement is likely to be profoundly influenced by their cultural background, I think there are certain aspects that are common to all of us.

I've noted before in this book that the study of the Universe gives us insights into the nature of space and time, into our origins and into the ultimate fate of our species. With its emphasis on the big questions, astronomy is not too far removed from such deep-seated matters as the nature of existence, the meaning of life and perhaps even notions of spirituality. I've already touched on some of these issues, but I hope you will allow me to guide you through a few more in this final chapter. First, though, let me add just a little more to the story of the Garnetts.

Not surprisingly, perhaps, my mother's experience of pregnancy and childbirth in wartime was the stuff of family legend. How could she forget one particular visit to the doctor when she had taken along the customary urine sample in her shopping basket? Wartime shortages meant that Mum had to make do with whatever container was available, and in this case it was a long-empty quarter-bottle of whisky. But when she turned up at the surgery, the bottle was gone—stolen from her basket. Oh, such poetic justice. And then, when I finally came into the world, it wasn't in the spotless surroundings of a maternity ward; it was in the family home—in Lidget Green's cheerfully named Cemetery Road. Maybe there was poetic justice in that, too—although it could have been worse. Necropolis Road was just up the hill.

Notwithstanding the arrival of her bonny bouncing boy to ease the family's grief, my mother drew the greatest solace from her religious belief. She was never a fanatic, but she maintained her Christian faith and values with quiet conviction throughout her life, attending both Methodist and Anglican churches at different times. Mum led a full and busy life, raising two boys in circumstances that were seldom easy. But she never doubted for one second that someday, somehow, she would be reunited with her beloved brother and the rest of her family. When she herself died, in 1992, it was with a serene confidence that the best was yet to come, and I envied her the simplicity of that unshakeable belief.

At the beginning of this book, I mentioned that it's safest to keep God out of discussions on astronomy, and that is not going to change now. You'll have to make up your own mind about God. Religion, however, being a human construct, is a different thing. And you've only to think of Galileo and the Inquisition to see astronomy and religion colliding head-on. Having written his *Dialogue on the Two Chief World Systems, Ptolemaic and Copernican*, a provocative book arguing that only a simpleton would refuse to believe that the Earth revolves around the Sun, Galileo fell foul of the Holy Roman Church. His case was not helped by his incautious identification of the simpleton with the pope, and, on 22 June 1633, he was convicted of vehement suspicion of heresy. Poor old Galileo. His punishment included a formal abjuration, prohibition of his *Dialogue*, imprisonment at the pleasure of the Inquisition, and religious penances. He was lucky to escape with his life—unlike the hapless Giordano Bruno, who had been burned at the stake in 1600 for maintaining similar beliefs.

Religion has come a long way since Galileo's time, although we all know that there's still a fanatical minority out there. But look at what the great religions of the world have given us. Our society's underpinning of ethics and law, its humanitarian values, its efforts to alleviate poverty and suffering—these are just the start. And one only has to think of the literature, art, architecture and music they have inspired to realise how much poorer our cultural heritage would be without them. And, yes, the world's religions have arisen as a result of the lives of a few extraordinary individuals—historical figures with profound insights, who undoubtedly had a momentous impact when they walked the Earth. When it is those individuals' quest for truth that is at issue, we might even see religion and science as different sides of the same coin. So, I have no problem with religion, as long as no one decides to make it their business to convert me to their own beliefs.

There are a couple of circumstances in which I do take issue with religion, however. The first is when it is used to the disadvantage of others, and, unfortunately, that is all too often. It's not just the dramatic events that are cases in point—the Kuta Beaches and Twin Towers of today's world. Nor is it whatever religious fervour drove the warrior-priests of Sechín, in Peru's Casma Valley, to butcher their rivals three millennia ago. It's not just sectarian violence in Afghanistan, and it's not just burnings at the stake in the Reformation. Sadly, we all know that child abuse, persecution, political intrigue, bigotry, fraud, moral scandal and countless other crimes and injustices have wormed their evil ways through organised religion over the centuries. Religion is not alone in that, of course; such evils can permeate all facets of human activity, regardless of the underlying belief system. But they are particularly prevalent when there are vulnerable individuals to be preyed upon—exactly the kinds of individuals for whom religion should be a protective shield. Corruption seems to have the potential to become endemic wherever religion holds sway.

Another area in which I believe religion is a bad thing is when it insists on people believing things we know not to be true. Once again, that is not exclusively the province of religion—one has only to think of the lies of certain totalitarian regimes, for example. But, with hindsight, it's essentially what happened to Galileo. His quest for truth, which was based on observation and reason, fell foul of the unchanging dogma of the Church. Something similar is happening today, particularly in certain parts of the United States, where the idea of an intelligent designer being responsible for *Homo sapiens* is held up as equal in scientific legitimacy to the idea of evolution by natural selection. By extension, the same confrontation includes cosmology. While intelligent design is based on ancient texts, cosmology and biological evolution come from the very best evidence-based research. And, unlike the ancient texts, in science nothing is set in stone—the emergence of new evidence will often bring about a change in our understanding. Though the ancient texts might well be the writings of a lot of wise people, those individuals were essentially trying to describe the world around them equipped with nothing more than their five senses. That is fine in a historical and cultural context, but can't legitimately be used as evidence against scientific research.

Occasionally, at one of my talks, I'm taken to task by a creationist—a believer in intelligent design. I always try to engage with such folk with as much sensitivity as I can. The way I put forward science's view is to ask when they last flew in a jet aircraft. It's usually quite recently, but that's not the point. The issue is that they boarded it at all. A modern aircraft is the sublime result of a century of scientific research in aerodynamics, mechanical engineering, materials science, electronics and so on. It is the epitome of evidence-based development. And exactly the same processes of observation and reason used in aeronautical engineering are what have led us to our views on evolution and cosmology. So why would you trust your life to a jet if you don't want to trust evolution and cosmology? Shouldn't you instead trust some sort of creationist technology to get you from Sydney to London? (It might take you a while . . .)

I have some sympathy with the plight of those who are told what they must believe. I, too, was once an enthusiastic participant in religious activity. For a few years in my youth, I belonged to an independent evangelical church that had modelled itself on the early Methodists, with healthy doses of John Wesley, Charles Haddon Spurgeon and Ira D. Sankey thrown in. Notwithstanding the occasional servings of hellfire and damnation that were dished out in the Sunday gospel meetings, I still feel some attachment to that period, perhaps because of the hormonal blur within which I was living at the time. Oh, all right, yes—and I was engaged to the minister's daughter.

Although, disquietingly, there seemed to be a lot of doctrinal in-fighting with other Christian groups (always conducted with due decorum), I'm glad to say that to the best of my knowledge, my church was free from the kind of corruption mentioned above. I would, however, question my belief at the time that it was my responsibility to evangelise the world—which I duly attempted to do. This was an effort to persuade people towards a particular religious viewpoint based on its imagined merits, which I would now consider unethical. It also led to what was probably the most ill-advised overseas trip of my life, when I went off trying to sell Bibles to the Italians. Best draw a veil over that one. As time went by, I began to question some of the basic foundations of the creed to which I adhered. While much of it was relatively self-consistent, there also seemed to be a lot of wishful thinking involved. Evidence of life-transforming experiences of deliverance could equally well be put down to a nurturing community, improved psychological well-being or simple coincidence. And, unlike the work of the Salvation Army—which I still admire—much of what we did seemed rather unrelated to the problems of the real world. Gradually, I drifted away—sadly, from both the church and the minister's daughter. My belief system changed. But I would say that the experience of having seen religion from the inside was extremely valuable, and perhaps even transforming.

LAST RITES?

Despite the pros and cons in this unashamedly naive assessment of religion, there is one facet of the human condition in which, in my view, religion reigns supreme. At the end of life, religion has no competitor in providing comfort both to the dying and to the bereaved. With the promise of an afterlife, individuals facing death, or those mourning a loved one's passing, have much by which to feel reassured. Within the Christian faith, Roman Catholics are masters at this. Their last rites are a solemn preparation for death, while the ritual framework that constitutes a Catholic funeral is perfect for giving the deceased an appropriate send-off. Everyone knows the rules, everyone knows what they're supposed to believe, and everyone knows what they have to say in praying for the soul of the departed.

Times are changing, however. A recent study of census data recorded over periods of up to a century has found a steady increase in the number of people who claim no religious affiliation. Undertaken by US academics, the study looked at data from nine developed nations, including Australia, New Zealand and Canada. It's perhaps not surprising that these modern secular democracies should show a decline in religious sentiment, but further mathematical analysis of the data indicated that in the countries concerned religion will probably fade away completely. The authors of the study have drawn a parallel with the way some of the world's lesser-spoken languages have declined in use and are slowly dying out.

At least one eminent author counsels against writing off religion too hastily. In his book *A Short History of Christianity*, Geoffrey Blainey notes that the Christian faith has undergone many ups and downs in popularity in its 2000-year history—and still survives. But the fact remains that it has never had to withstand an educated onslaught from the secular world of the kind taking place today. It seems that the decline is occurring throughout the whole of the western world's Judaeo-Christian strongholds, and perhaps elsewhere, too.

Against this backdrop of declining religious belief, what is waiting in the wings to provide comfort at the end of life? In recent years, there has been an increase in secular funerals, which provide an opportunity to celebrate the life of the departed with due dignity. However, they still gloss over the big question: what happens to you when you die?

Science's answer is pretty well summed up by theoretical physicist Stephen Hawking's widely publicised comments on death: 'I regard the brain as a computer which will stop working when its components fail. There is no heaven or afterlife for broken down computers; that is a fairy story for people afraid of the dark.' While I rather take issue with the condescending tone at the end of this statement, I agree with the basic sentiment. When you die, that's it. The end. And it's no use pointing to near-death experiences as evidence of an afterlife. Today's best research shows that these are almost certainly a combination of physiological and psychological phenomena associated with the trauma being suffered by a dying brain.

I do think, however, that there's more to this issue than a mere switch-off at the end of life. As we saw in the last chapter, death plays an integral part in the natural order of things, in many ways foreshadowing rebirth. And who knows how new ideas like string theory and the multiverse might inform our view of death?

ABOUT TIME

Provocative as the idea of the multiverse is, it's not likely to provide much comfort to the bereaved. The idea of departed loved ones flitting off into an alternative universe is as much a fairy story as Hawking's afterlife for broken-down computers.

Much of the science we have discussed in the last couple of chapters is at the very frontier of knowledge. There are no certainties here, and we are surrounded by mysteries that seem almost overwhelming, suggesting we are really only scratching the surface of understanding, struggling with the vastness of our ignorance. As Martin Rees put it some years ago, 'What we have traditionally called the Universe may be just an infinitesimal part of reality.' But the role of science is to discover what we can about the true nature of that reality—and so we will continue scratching.

It seems to me that one of the greatest mysteries that faces physicists and philosophers alike also has the biggest potential to provide insights into death and bereavement. That is the nature of time—the ticking backdrop against which we live out our lives, which most of us take completely for granted, and which none of us can cheat. We can never make time go backwards. But we already know that time is interwoven with three-dimensional space in a single entity that has been called 'space-time' since the work of Minkowski in 1908. And we also know that every observer sees time differently, depending on their state of motion and the gravitational field they are experiencing. Time, therefore, is bendy—malleable, if you like. In fact, by undertaking some rather dramatic spaceflights (admittedly using rockets that haven't yet been developed), we could even travel forwards significantly in time relative to our stay-at-home colleagues. We would make use of the time dilation effect of special relativity, using our near-light speed to slow down our clocks relative to theirs. On our return, we would effectively have travelled into their future. It would be a one-way trip, though, because backward time travel is impossible by this method.

Relativity tells us that time is a dimension like space, but there is clearly a difference. We can move around in space, but, apart from the kind of one-way trip mentioned above, we can't move around in time. Moreover, while we perceive space all around us, we experience time only moment by moment. And it only goes in one direction, despite the fact that many of the processes of physics are actually time-reversible. The direction of the arrow of time seems to be set by such things as the inexorable increase in disorder throughout the Universe (which physicists call the Second Law of Thermodynamics) and the way radiation—light, for example—travels only forwards in time. As I've mentioned already, based on Einstein's general relativity, the standard model of the Universe requires time to have kicked off with the Big Bang. But should string theory or M-theory prove to have any validity, our ideas would have to change. These theories require time to have been ticking away before the Big Bang occurred, a notion that Einstein himself actually found more palatable.

Einstein also held the view that 'time is nothing but a stubbornly persistent illusion'. In this he was presaging the sentiment of a number of writers (most recently the physicist John Archibald Wheeler) that 'time is what stops everything happening at once', the implication being that perhaps the temporal contents of the fourth dimension somehow exist in their entirety. None of us doubts that all of space exists, so why shouldn't all of time exist, too, since the two are so intimately linked? If this were the case, and our moment-by-moment view of time is just some weird cross-section of a much deeper reality, it would be a truly remarkable thing. I have to say I find this both disturbing and heartening. Disturbing because everything we have ever done—good and bad—is still going on in some deep underlying reality. Somewhere, I'm still trying to sell Bibles in Italy, for example. And, not far away, Galileo is still languishing under house arrest. Crumbs. But it's heartening because it means those we have lost are still part of reality. Alongside our former selves.

Unfortunately, this is not something that lends itself to easy investigation. But while it would be utterly premature to suggest that the illusory nature of time is anywhere near being confirmed, support for the idea is coming from an unlikely source. And if you thought flexible space-time was strange, wait until you hear this.

TIME OUT

Relativity deals with the Universe on the scale of large things. At the other end of the range, however, is a quite different theory, which, as I have noted before, seems to be incompatible with relativity. Its name is probably the most misused scientific word in the English language (and several other languages, too), especially when it's commandeered to hype the marvellous properties of the latest quack remedy. But, for the better part of a century physicists have understood the behaviour of atomic particles in terms of quantum mechanics, which says that their physical parameters are quantised—that is, they change in tiny, fixed steps, rather than smoothly and continuously.

Now, you would be well within your rights to grumble that introducing quantum theory a few pages from the end of a science book is a bit like introducing Beethoven in the appendix of a book on classical music, or Greg Norman as a footnote in a book of golfing greats. I have to apologise for that. But there's probably enough space left to highlight the aspects of the theory that inform the idea of time as an illusion.

Quantum theory allows objects to behave very strangely indeed. They can be in several different states at once—upside down, right way up, moving, stationary, whatever—or even in several different places at once. They are said to be in a state of 'superposition', a word that comes from the similarly odd fact that such objects can equally well be described as waves, which can be superimposed on one another. All this unlikely behaviour is confirmed by experiment. But the experiments have to be done very cleverly—because the catch is that the act of observing quantum objects destroys their superposition, and their behaviour becomes perfectly 'normal'. That metamorphosis to a normal state is known by the exceptionally unattractive name of 'decoherence'. The idea of introducing the observer into the mix—an observer that could just as well be an electronic camera as a person—flies in the face of the objectivity of scientific investigations, which are supposed to be independent of the observer. But what is really happening is that objects exhibit weird quantum behaviour only when they are completely isolated from their surroundings. As soon as that isolation is broken—for example, by objects being illuminated so they can be observed—they settle down to just one state or position. It's because this isolation is much harder to achieve for large objects than small ones that quantum behaviour is usually associated with atoms or subatomic particles—and it's why we humans can't be in two places at once. Not yet, at least.

A number of laboratories are now observing quantum behaviour in steadily larger objects by carefully isolating them from their surroundings. A microscopic tuning fork, for example, has recently been observed to be simultaneously vibrating and not vibrating. And, more intriguingly, there is a growing recognition of similar processes taking place within living organisms in a new field known as 'quantum biology'. Remarkably, such disparate phenomena as photosynthesis in plants, magnetic sensing in migrating birds and our own sense of smell might have their origins in quantum effects. It's even possible that quantum computing (an active area of research because of its enormous potential in technology) is already taking place inside our own brains. This exciting field of study is sure to develop rapidly in coming years.

But there's one final level of weirdness in quantum mechanics that I want to mention, and that's something called 'entanglement'. This is a concept that Einstein hated—even though he actually first raised the idea, in 1933. 'Spooky actions at a distance' is what he called it. It refers to a mysterious connection that pairs of quantum particles can have. They must somehow be associated with one another—for example, pairs of light particles, or photons, that have been fired from the same laser. In the quantum world, these entangled particles behave like a single entity. When one of them loses its state of superposition and starts behaving like a normal particle, so does the other. And, bizarrely, it doesn't matter if the two entangled particles are right next to one another or half a Universe apart; they both lose their superposition simultaneously. So if you know what state one entangled particle is in (which way it's spinning, for example), then you also know the state of the other—because they're always complementary. This phenomenon, which is known as 'nonlocality', seems to break all the rules. How can one atom or photon know what another is doing when it would require faster-than-light travel for the information to be passed between them? But it has been demonstrated experimentally many times since the 1970s and, indeed, is now at the heart of a signal-processing technology called 'quantum encryption'. Entanglement provides a serious hint of an underlying reality that doesn't obey normal physics, especially when we see weird quantum effects being demonstrated in the ever-bigger assemblages of atoms that constitute normal-sized objects.

One of today's leading experts on entanglement, physicist Vlatko Vedral, thinks that at the deepest level neither time nor space exists as a real entity. Vedral writes that 'the entanglements . . . interconnect quantum systems without reference to space or time . . . We must explain space and time as somehow emerging from fundamentally spaceless and timeless physics.' He suggests that normal space-time emerges out of quantum entanglements through the process of decoherence. While that may sound dauntingly complex, to me it suggests a view of the Universe that is once again deeply heartening, particularly for those of us confronting human mortality in some way. If neither time nor space exists at the most fundamental level, it means that all our everyday tribulations—our successes and failures, gains and losses—are mere shadows of reality. A good analogy might be the fact that seemingly solid objects like bricks and iron bars are really just collections of atoms with a whole lot of empty space in between them. That doesn't mean you could slide your way out of gaol if you happened to be there—but it would be a reassuring thought.

It may well be that our chances of peeling away the layers of mystery surrounding such 'spaceless and timeless physics' are slender—for the present, at least. As the physicist Max Planck, one of the founding fathers of quantum theory, commented, 'Science cannot solve the ultimate mystery of Nature, and it is because in the last analysis, we ourselves are part of the mystery we are trying to solve.' Planck was no doubt alluding to the effect of an observer in rendering a quantum object decoherent, but there is a wider truth in his remark. And it brings me to one final aspect of science's place in our deepest understanding of the Universe.

The historian Geoffrey Blainey, whom we met a few pages ago, has lamented the decline of wisdom as a human virtue in the face of knowledge. I freely admit that the quest for knowledge—rather than wisdom—is what has motivated me for the past 40-odd years. But I would submit that this quest brings with it a wisdom of its own. It comes from a sense of wonder at the splendour and majesty of the Universe in all its complexity, and from the fact that we are part of it. It comes from a profound sense of humility about our true place in the Universe, together with a feeling of awe that evolution has equipped each one of us with the most complex entity known in that Universe—the human mind. It comes from an overwhelming sense of privilege that those minds can seek to comprehend such mysteries, particularly in the era in which we live, when science is exploding with so many new discoveries that a book like this becomes out of date almost before it's published. And it comes from unbridled excitement and optimism about what we have yet to find: the great unknowns that could transform our understanding of absolutely everything, including death and bereavement.

For me, the litmus test as to whether science really can offer anything to the bereaved is to imagine telling my grandfather about the possibility of deep underlying truths concerning space and time. As a gruff and grumpy Yorkshireman with his feet firmly on the ground, he was not one to suffer fools gladly—especially me. But he had wisdom, too, born of a different kind of hope, and I think I can guess what he would have said. 'Aye, lad, I know. And that's why I go to t'church.' And who would I be to advise him to do anything different?

EPILOGUE

The skies of Lyngenfjord weren't quite as dark as we had expected. Though remote from major light pollution, its shores were dotted with tiny fishing villages whose street lamps were visible as twinkling rows of orange light reflected in the icy water. Rather charmingly, I have to admit. And the faint glow of Tromsø, 60 kilometres to the west, was discernible over the peaks of the Lyngen Alps across the fjord. But none of that mattered. Not even the Moon, skipping over the mountain tops from one night to the next as it waxed from a crescent to first quarter and beyond, could dim nature's brilliance.

We soon discovered that there was a pattern to the auroral displays. The fun would start early in the night, shortly after the long twilight had ended, with a hint of a glow on the northern horizon. Then, around 8 pm or so, we'd detect a thin greenish band snaking from east to west through the far northern constellations that graced the skies of Lyngenfjord. The Plough, Cassiopeia, Ursa Minor—star-patterns that were unfamiliar to those of us more accustomed to the Southern Cross, Vela and Centaurus. As we watched, the green band would broaden, and we would begin to see delicately structured filaments of light stretching upwards, with more bands starting to fill the patches of dark sky.

And then, if we were lucky, the show would ramp up to its evening crescendo. Swirling curtains of light, twisted into impossible shapes, would sweep in waves across the sky, taking only seconds to form, brighten into prominence and then fade again. Still more green bands would roam from east to west. And this silent display would be interrupted by our shouts of excitement, as we looked on in awe. By now, our fumblings with tripods, ISO settings and exposure times had been transformed into a model of proficiency. With surprising deftness, we performed the nightly balancing act of getting enough light into our cameras while still managing to capture the aurora's fleeting patterns. And all the while trying not to slip over on the icy snow, or expose our fingers and thumbs too long to the frozen air. The only question was where to point our cameras, as the sky exploded into curtains of green, tinged above with red, and below with a pale magenta.

Sometimes, these shows put on a rare climax, with the formation—almost overhead—of an auroral corona. Like crystals precipitating out of a super-saturated chemical solution, green, finger-like rays would burst from the zenith before our astonished eyes. We knew this was a trick of perspective, caused by parallel columns of light forming along the near-vertical lines of the Earth's magnetic field, but that made it no less extraordinary. Except in the most extreme conditions of auroral activity, this display is exclusively the province of those who venture to the planet's polar regions. This week, that meant us—for our vantage point in far-northern Norway was well inside the Arctic Circle. And on those frozen, magical nights early in 2012, members of the Fire in the Sky study tour from Down Under wouldn't have wanted to be anywhere else in the world.

In earlier times, the Sami people who inhabited this frostbitten country regarded the Aurora Borealis with fearful awe. Legends about the origin of these eerie lights were intertwined with traditions concerning the souls of the departed, and their campfires flickering mysteriously beyond the northern horizon. The lore of the dancing lights dictated that you must remain as inconspicuous as possible in their presence, lest you attract their baleful gaze. Wearing white was out of the question. In western European culture, too, the Northern Lights belonged in the realm of mystery. A broadsheet printed in Augsburg in 1570, and now in the Crawford Collection of the Royal Observatory, Edinburgh, depicts a bright auroral display that occurred on 12 January of that year as a row of heavenly candles above the clouds. No doubt the artist was at a loss to imagine how else such an unearthly apparition could be represented.

It was only at the turn of the twentieth century that the first glimmer of understanding began to emerge as to the true cause of the aurora. But it was an idea so outlandish, and so unpalatable to the scientific establishment of the time, that it was rejected almost out of hand, particularly in Britain. It generated a human saga of epic proportions—and it had its origins not far from the small Norwegian town of Alta, where our study tour had begun its rendezvous with the dazzling aurorae of 2012.

Just to the west of Alta, in the spine of mountains that dominates Norway's far-northern coastline, a high plateau called Haldde became the site of the world's first auroral observatory. It was set up by a visionary Oslo University scientist called Kristian Olaf Birkeland, who had a wild notion that the Aurora Borealis was caused by subatomic particles in space, and was somehow linked with the Earth's magnetic field. To this end, Birkeland and a handful of colleagues wintered on two mountain peaks at Haldde during the bleak months of darkness from late 1899 to early 1900.

Measurements of the local magnetic field were correlated with appearances of the lights, and attempts were made to photograph the auroral displays simultaneously from the two peaks—Sukkertop and Talviktop—in order to estimate their height by triangulation. This was intended to reveal whether or not the aurora actually touched the ground, since many scientists still believed it was purely a meteorological phenomenon. In his first expedition, Birkeland failed to make these measurements due to the insensitivity of his photographic equipment, but observations a few years later proved that aurorae do, indeed, populate the upper atmosphere at levels above 90 kilometres.

It was Birkeland's formulation of a radical theory in the wake of that early work that made him so unpopular with British scientists. He postulated that the Sun was a source of the subatomic particles that were then known as cathode rays (but are today called electrons), and that their interaction with the Earth's magnetic field near the poles caused them to excite the atoms of the upper atmosphere into a frenzy of luminescence. He later suggested that the particles we now call protons might also be emitted from the Sun, and play their part in the celestial light-show.

The idea that the Sun could be responsible for anything other than heat, light and gravity was what so incensed Birkeland's critics. And Britain's Royal Society led the charge, believing that its scientific heritage effectively endowed it with ownership of these phenomena. After all, they had been the province of Newton and Herschel, among others. But subatomic particles from the Sun? Proposed by a *Norwegian*? Oh, dear me, no.

As the twentieth century dawned and matured into the gentility of the Edwardian era, Birkeland's ideas were put to the test at the University of Oslo. In his Terrella ('little Earth'), he set up a magnetised model of the Earth, enclosed in a vacuum chamber, which he could bombard with electrons. Sure enough, a recognisable simulation of the aurora was generated. Locally, at least, Birkeland gained credibility, and was encouraged by the university authorities to continue his relentless pursuit of the aurora's mechanism.

But with the drums of war now beating loudly on the horizon, he also turned his attention to other inventions. He invested considerable time and effort in the development of a novel electric cannon, at the same time following more peaceful ambitions in perfecting an electric furnace for the fixation of atmospheric nitrogen for fertiliser. That work led to the creation of the Norsk Hydro company—still a major industrial force in aluminium and renewable energy.

Birkeland would have been the first to admit that these diversions were undertaken merely to allow him to fund his obsession with pure research. As the First World War lumbered forth on its dreadful course, his work took him to Egypt to investigate a phenomenon called the zodiacal light, a faint luminous band that towers over the horizon at dusk and dawn. He wanted to discover whether this, too, had an electromagnetic origin. Sadly, though, amidst growing paranoia about being persecuted by the British authorities, Birkeland began to sink into a mire of mental instability from which he never recovered. His condition led to growing usage of sleeping preparations, which eventually reached dangerous levels. In 1917, en route homewards from Egypt via Japan to avoid the British, Birkeland died in Tokyo at the age of only 49.

With Birkeland gone, there was no longer a champion for his ideas, and the notion of electric currents carried through space by a solar wind of particles dropped off the agenda in the face of a hostile, principally British, scientific establishment. The origin of the polar aurorae was relegated to the too-hard basket. But half a century after his death, when the space age was still in its first flush of youth, magnetic measurements taken from Earth-orbit revealed what ground-based observations had failed to detect: that space is full of subatomic particles—electrons *and* protons. Astonishingly, on both counts, Birkeland had been right.

Not quite so insightful was his research into the zodiacal light. This is now known to originate in the flattened ring of dust particles surrounding the Sun, which envelops the inner planets. While the streams of electrons and protons from the Sun certainly buffet the dust particles, they don't play any part in exciting them to luminosity, as they do with atoms in the Earth's upper atmosphere. Rather, the zodiacal light is caused simply by the reflection and scattering of sunlight.

Aurorae, on the other hand, are the result of a highly complex interaction between the wind of particles from the Sun and the Earth's magnetic field. Our modern understanding includes such subtleties as the occurrence of aurorae in a circular zone around each magnetic pole rather than a concentration of light at the pole itself—something that was not explained by Birkeland's theory. We also know that aurorae occur on Jupiter, Saturn, Uranus and Neptune, all of which have their own magnetic fields. And our knowledge of the energies carried by the solar particles lets us understand why we see aurorae of different colours at different heights. The prominent green bands occur between 100 and 200 kilometres above the Earth, and are caused by atmospheric oxygen atoms being excited. Often, there are extended pillars of red light above them, which are again due to oxygen atoms, but at a lower energy. In the most energetic displays, molecules of nitrogen are excited below 100 kilometres, causing them to emit red, blue and violet light, giving a characteristic magenta fringe to the underside of bright aurorae.

We know, too, that the solar wind of particles is variable in its intensity. The importance of this cosmic buffeting of the Earth's environment is so great that we now refer to it as 'space weather', and monitor it with a flotilla of robotic spacecraft. In fact, the solar wind has its origin in magnetic activity taking place near the surface of the Sun. As our star progresses through the eleven-year cycle that was discovered in the 1840s by observing variations in the number of sunspots on the Sun's surface, it becomes increasingly prone to the occurrence of magnetic storms in its atmosphere. These storms propel energetic particles and magnetic fields into space at up to 1200 kilometres per second, and generate powerful shockwaves in the solar wind. They can damage orbiting satellites and disrupt power distribution grids. The underlying mechanism of the storms is now known to be the sudden release of magnetic stress in the Sun—a kind of gigantic magnetic 'twang'—and the end-product is a monumental dumping of energy into the inner Solar System. When this occurs, aurorae become visible nearer the equator than normal, and ground-level magnetism increases dramatically—as was noted by Birkeland. These phenomena tend to occur near the maximum of the Sun's eleven-year cycle of sunspot activity, and each event typically lasts for a day or two. Such is the state of our knowledge, only a century or so after Birkeland pioneered the discipline of space physics.

Photographs taken in Oslo while Birkeland was experimenting with his Terrella show him wearing a fez. Although that was not uncommon among scientists of his era, I've always thought of it as a bad sign, an indication that someone could be on the slippery slope to madness. If you ever see me wearing a fez, shoot me. For Birkeland, what had begun as a passionate interest ended as a maniacal obsession. Perhaps because of that, history has consistently undervalued his contribution. While he is celebrated on the Norwegian 200 kroner note, as well as in museums at Alta and elsewhere, he is not well known outside his own country, even today.

Our Fire in the Sky tour in January and February 2012 paid due homage to Birkeland. But what dictated its timing was, of course, the Sun. Predicted to reach a peak in its cycle during 2012 and 2013, the Sun was likely to be more active than it had been for almost a decade, with the promise of enhanced auroral activity. And it didn't disappoint us. Blessed with clear skies, we were able to experience for ourselves this most awe-inspiring celestial phenomenon, rivalled in grandeur only by a total solar eclipse. For a brief spell, we allowed ourselves to share the apprehensions of the Sami, while experiencing a face-to-face encounter with the magnetic workings of our planet using state-of-the-art photographic technology to document the astonishing colours that lit the sky. We left the Arctic enraptured.

But, as I've mentioned before, Marnie has a flair for putting together itineraries that cater for everyone on our study tours. Thus it was that after our rendezvous with the Northern Lights we embarked on an expedition to other attractions of the region, both scientific and otherwise. Alta and Lyngenfjord are within striking distance of Narvik, for example, which boasts a spectacular railway line linking the famous port with Kiruna in northern Sweden. And Kiruna has a rocket range—the Esrange, operated by the Swedish Space Agency, whose restricted airspace will eventually be used by Richard Branson in connection with his Spaceport Europe. From Kiruna Airport, Virgin Galactic will fly well-heeled passengers *through* the Aurora Borealis, as opposed to their simply observing it from below. This is perhaps the most intimate encounter possible.

Southern Scandinavia, too, has much of interest. Not far from Stockholm is Kvistaberg, where the historic Uppsala University has custody of one of the largest Schmidt telescopes in the northern hemisphere. With an aperture of 1 metre, the Kvistaberg Schmidt bears a distinct family resemblance to its larger siblings, the 1.2-metre Oschin Schmidt Telescope at Palomar Mountain, and our own UKST at Siding Spring Observatory.

Although equipped with a modern electronic camera, the telescope is little used today due to indifferent weather conditions compared with overseas sites. But the astronomical jewel of southern Sweden is a place we have already visited in this book—Tycho Brahe's island of Ven in the Øresund. In autumn the island is enchanting, but a winter mantle of snow and the occasional basking seal on the Øresund's ice-floes made our February visits just as magical.

And so on. There's little space here to mention our trip to Tartu in Estonia to see telescope pioneer Joseph Fraunhofer's masterpiece: his 24-centimetre Great Dorpat Refractor of 1824, which set the pattern for refracting telescopes throughout the nineteenth century. Nor our visit to Tuorla Observatory in Finland, where the tall tower housing its 1-metre reflecting telescope of 1959 sprouts incongruously from a snow-laden pine forest. Nor a captivating night-time excursion to Copenhagen's curious Round Tower of 1642, whose 209-metre-long internal spiral ramp allowed horse-drawn carriages to access an astronomical observing platform high above the streets of the city. Oh, and did I tell you about Iceland, with its volcanoes, hot springs and glaciers . . . ?

Besides science and history, there is one other strand to our touring. It arises from a lifelong passion of mine for the great music of the world, and it's what took our Stargazers to hear the Berlin Philharmonic Orchestra in Chapter 1. In the Fire in the Sky tour, it took us to the Sibelius Museum in Turku, to celebrate the life and work of the great Finnish composer. But it also resulted in an encounter that was, in its own way, as baffling as my meeting with Messenger Nine. What it lacked in comprehension, however, it more than made up for in inspiration.

Urmas Sisask is an Estonian composer, very well known in his own country, who writes music celebrating the Universe. Intrigued by the astronomical connection, Marnie had attempted to contact him while planning Fire in the Sky, in the hope that our tour participants might catch a concert of his music. Those attempts came to nothing, perhaps because of Sisask's unfamiliarity with English. Then, by a remarkable coincidence, she was approached by the Griffyn Ensemble, a Canberra chamber group, wanting to know whether I might be interested in narrating a series of Australian recitals of the music of—Urmas Sisask. The recitals did, indeed, take place after our tour, but the approach from Griffyn opened the door to an extraordinary experience in Estonia.

Sisask's fame is such that when our local guide, Katarina, was told that she would accompany us on the drive from Tallinn to his conservatorium in the village of Jäneda—and act as his interpreter—she wept with joy. That was the first surprise. The second was Jäneda itself. As we forsook Tallinn's stunning mediaeval centre for its drab, Cold-War era suburbs, we had low expectations of the snow-covered landscape beyond. But ugly apartment blocks and power stations gradually gave way to pine forests and charming villages. And Jäneda was pure Christmas-card delight. At the heart of the village is the imposing manor, built between 1913 and 1915, with a history that links its first occupants, Johan von Benckendorf and Maria Zakrevskaya Benckendorf, to the Russian Revolution—and Maria to a life of intrigue and espionage. Amazing stuff. But that wasn't why we were there.

When we entered the manor house, which now doubles as a conference centre, we were met by Sisask himself, and ushered up a long staircase to a shuttered room. In the centre was a piano surrounded by chairs, and all around the lower walls were pictures that looked as if they had been removed from old astronomy magazines. Outdated they might have been, but they included all the familiar iconic images of stars, galaxies, nebulae and planets made with the Hubble Space Telescope and other great telescopes of the world. They betrayed a humbling passion for the sky, pursued with minimal resources.

With the study tour members comfortably seated, and Katarina interpreting, Sisask explained that we were now sitting in his 'astromusical observatory'. He apologised for the run-down state of the decor, and the stains where the roof had leaked, but made it clear that this was a very special place to him. None of us had ever heard of an astromusical observatory, so this was all rather intriguing. To be honest, I could sense that some of our party were wondering what on Earth they had got themselves into. Then, with Katarina struggling to translate the more technical aspects of Sisask's introductions to his works, he launched into a spectacular pianoforte tour through his musical universe. His music is wide-ranging in style and accessibility: some pieces are almost impenetrable, while others you can hum along to—and then can't get out of your head. Unexpectedly, I found parallels with the music of Australian composer Ross Edwards. Perhaps that is not so surprising, given that both men are deeply influenced by the natural world. Sisask performed his work with passion and commitment, and showed great generosity as he explained what each piece was about. OK, he acknowledged, the piano was a bit out of tune—perhaps because of the damp—but that was just an integral part of the performance. Fair enough, I suppose. When resources are scarce, it helps to be able to rationalise. And, as the recital progressed, his audience warmed in their mood, from polite bemusement to clear delight, especially when Sisask introduced some of his better-known works.

His finale was the undoubted *tour de force*. Beginning with a repeated pattern of notes on the piano, he quickly delegated the piano-playing to his wife, while he embarked on a virtuosic tour of the world with a dozen or more folk instruments from different countries, which he played in rapid succession. As this thrilling piece evolved into a frenzied celebration of life, the listeners responded with enthusiasm—especially when Sisask produced a didjeridu, and showed himself to be no mean exponent of circular breathing. But no one was prepared for the climax. With his wife pounding away, and Sisask quickly switching from one instrument to the next, we suddenly became aware that the family dog had become part of the performance, clomping up and down the keyboard to make its own contribution to the melody. Jaws dropped around the room. While none of us truly understood this strange music—except, perhaps, the dog—Sisask won all our hearts with the performance. The recital ended with thunderous applause and warm congratulations—and the handful of CDs he had available for sale were snapped up in a twinkle. Clearly, in Sisask's resource-starved world, every eurocent counted.

But this remarkable man had one more trick up his sleeve. And it was pure magic. Dimming the lights of the room, he revealed what he had meant by an astromusical observatory. As we grew accustomed to the darkness, we were astonished to discover that the walls and ceiling were covered in stars. With the lights on, they had been invisible against the flaking paint, but now, boosted into fluorescence by a hidden ultraviolet lamp, they shone in all the profusion of an arctic night. Yes, they were just stick-on stars of the kind you'd buy for a kid's bedroom, but this was no haphazard decoration of a would-be astronomer's den. All the constellations of the northern sky—the Plough, Cassiopeia, Ursa Minor, and the rest—were clearly recognisable in Sisask's humble recital room. It was, in effect, a planetarium.

Not only had the stars been placed with painstaking accuracy in regard to their positions, but also their relative brightness was correct. This was a labour of love of monumental proportions, and Sisask told us that he had positioned no fewer than 2200 stars to make the observatory. That's two-thirds of the real stars visible to the unaided eye on a clear, moonless night. To witness the care and dedication that this musical genius had bestowed on his fascination with astronomy was inspiring.

Emerging from the manor house into the weak late-afternoon sunshine, I felt I had encountered a kindred spirit. The language barrier didn't matter. A blend of astronomy and music had permeated his life, and had propelled his creativity in a unique direction. And he had brought enormous pleasure and insight to the music-lovers of Estonia. His fame was well deserved. But, as we boarded our coach back to Tallinn, I couldn't help reflecting that sticking on all those stars must have almost driven him mad. Star-craving mad.

FURTHER READING

Aughton, Peter 2004, *The Transit of Venus: The Brief, Brilliant Life of Jeremiah Horrocks, Father of British Astronomy*, Weidenfeld & Nicolson, London.

Blainey, Geoffrey 2011, *A Short History of Christianity*, Penguin, Melbourne.

Christianson, John R 2000, *On Tycho's Island: Tycho Brahe and His Assistants, 1570–1601*, Cambridge University Press, Cambridge.

Copernicus, Nicolaus 2004 [1543], *On the Revolutions of the Heavenly Spheres*, Running Press, Philadelphia, PA.

Couper, Heather and Pelham, David 1985, *The Universe: A Three-dimensional Study*, Random House, New York.

deGrasse Tyson, Neil 2009, *The Pluto Files: The Rise and Fall of America's Favorite Planet*, WW Norton, New York.

della Porta, Giovanbaptista 1589, *Magia naturalis* [Natural Magic], GG Carlino and A. Pace, Naples.

della Porta, Giovanbaptista 1593, *De refractioni* [On Refraction], GG Carlino and A. Pace, Naples.

de Waard, Cornelis 1906, *Invention of the Telescope*, American Philosophical Society, The Hague.

Einstein, Albert 1961, *Relativity: The Special and the General Theory*, Three Rivers Press, New York.

Frame, Tom and Faulkner, Don 2003, *Stromlo: An Australian Observatory*, Allen & Unwin, Sydney.

Freeman, Ken and McNamara, Geoff 2006, *In Search of Dark Matter*, Springer-Praxis, Chichester.

Galilei, Galileo 1610, *Sidereus Nuncius* [The Starry Messenger], trans. Albert Van Helden, n.p., Chicago.

Gamow, George 1965, *Mr Tompkins in Paperback*, Cambridge University Press, Cambridge.

Gascoigne, SCB, Proust, KM and Robins, MO 1990, *The Creation of the Anglo-Australian Observatory*, Cambridge University Press, Melbourne.

Ghezzi, Iván 2006, 'Religious Warfare at Chankillo', in W Isbell and H Silverman (eds), *Andean Archaeology III: North and South*, Springer, New York, pp. 67–84.

Ghezzi, Iván and Ruggles, Clive 2007, 'Chankillo: A 2300-year-old Solar Observatory in Coastal Peru', *Science*, vol. 315, pp. 1239–43.

Glass, Ian S 2008, *Revolutionaries of the Cosmos: The Astro-Physicists*, Oxford University Press, Oxford.

Gratzer, Walter 2002, *Eurekas and Euphorias: The Oxford Book of Scientific Anecdotes*, Oxford University Press, Oxford.

Gregory, James 1663, *Optica Promota* [The Advance of Optics], n.p.

Hawking, Stephen with Mlodinov, Leonard 2005, *A Briefer History of Time*, Bantam, London.

Hearnshaw, JB 1986, *The Analysis of Starlight: One Hundred and Fifty Years of Astronomical Spectroscopy*, Cambridge University Press, Cambridge.

Jago, Lucy 2001, *The Northern Lights*, Hamish Hamilton, London.

King, Henry C 1955, *The History of the Telescope*, Griffin, London.

Lomb, Nick 2011, *Transit of Venus: 1631 to the Present*, Powerhouse Museum, Sydney.

Machicado Figueroa, Juan Carlos 2002, *When the Stones Speak: Inka Architecture and Spirituality in the Andes*, Inka 2000 Productions, Cusco.

Malin, David 1999, *The Invisible Universe*, Little, Brown, London.

McEvoy, JP and Zarate, Oscar 1996, *Introducing Quantum Theory*, Icon, Cambridge.

Mizon, Bob 2002, *Light Pollution: Responses and Remedies*, Springer, London.

Newton, Sir Isaac 1687, *Philosophiae Naturalis Principia Mathematica* [The Mathematical Principles of Natural Philosophy] (known as the *Principia*), Macmillan, London.

Pepin, M Barlow 2004, *The Emergence of the Telescope: Janssen, Lipperhey and the Unknown Man*, rev edn, T Tauri Productions, Duncanville, TX.

Rees, Martin 1999, *Just Six Numbers: The Deep Forces that Shape the Universe*, Basic Books, New York.

Reiche, Maria 1968, *Mystery on the Desert*, published privately by the author, Hohenpeißenberg.

Robinson, Andrew 2005, *Einstein: A Hundred Years of Relativity*, ABC Books, Sydney.

Ruggles, Clive LN (ed.) 2011, *Archaeoastronomy and Ethnoastronomy: Building Bridges Between Cultures*, International Astronomical Union Symposium 278, Cambridge University Press, Cambridge.

Russell, Henry N, Dugan, Raymond S and Stewart, John Q 1926 (Vol. I) and 1927 (Vol. II), *Astronomy*, Ginn, Boston.

Tatarsky, Daniel 2010, *Dan Dare, Pilot of the Future: A Biography*, Orion, London.

Thoren, Victor E (with contributions by John R. Christianson) 1990, *The Lord of Uraniborg: A Biography of Tycho Brahe*, Cambridge.

Turnbull, HW (ed.) 1959, *The Correspondence of Isaac Newton, Vol. 1: 1661–1675*, Cambridge.

Van Helden, Albert 1977, 'The Invention of the Telescope', *Transactions of the American Philosophical Society*, vol. 67, part 4, n.p.

Watson, Fred 2004, *Stargazer: The Life and Times of the Telescope*, Allen & Unwin, Sydney.

Watson, Fred, 2007, *Why is Uranus Upside Down? And Other Questions About the Universe*, Allen & Unwin, Sydney.

Watson, Fred (chief consultant) et al. 2007, *Universe*, Millennium House, Sydney.

Webb, Stephen 2004, *Out of This World: Colliding Universes, Branes, Strings and Other Wild Ideas of Modern Physics*, Praxis-Copernicus, New York.

Webb, Stephen 2012, *New Eyes on the Universe: Twelve Cosmic Mysteries and the Tools We Need to Solve Them*, Springer-Praxis, Chichester.

Zinner, Ernst 1988 [1943], *The Origins and Dissemination of the Copernican Doctrine*, Verlag CH Beck, Munich.

INDEX